cture Notes in Computer Science

7: Fundamentals of Computation Theory. Proceedings, 1981. by F. Gécseg. XI, 471 pages. 1981.

8: Mathematical Foundations of Computer Science 1981. edings, 1981. Edited by J. Gruska and M. Chytil. XI, 589 pages.

9: G. Hirst, Anaphora in Natural Language Understanding: vey. XIII, 128 pages. 1981.

20: L. B. Rall, Automatic Differentiation: Techniques and Applis. VIII, 165 pages. 1981.

21: Z. Zlatev, J. Wasniewski, and K. Schaumburg, Y12M So- of Large and Sparse Systems of Linear Algebraic Equations. 3 pages. 1981.

2: Algorithms in Modern Mathematics and Computer Science. edings, 1979. Edited by A. P. Ershov and D. E. Knuth. XI, 487 . 1981.

3: Trends in Information Processing Systems. Proceedings, Edited by A. J. W. Duijvestijn and P. C. Lockemann. XI, 349 . 1981.

24: W. Polak, Compiler Specification and Verification. XIII, ages. 1981.

25: Logic of Programs. Proceedings, 1979. Edited by E. er. V, 245 pages. 1981.

6: Microcomputer System Design. Proceedings, 1981. Edited J. Flynn, N. R. Harris, and D. P. McCarthy. VII, 397 pages. 1982.

127: Y.Wallach, Alternating Sequential/Parallel Processing. 9 pages. 1982.

28: P. Branquart, G. Louis, P. Wodon, An Analytical Descrip- CHILL, the CCITT High Level Language. VI, 277 pages. 1982.

29: B. T. Hailpern, Verifying Concurrent Processes Using oral Logic. VIII, 208 pages. 1982.

30: R. Goldblatt, Axiomatising the Logic of Computer Program- XI, 304 pages. 1982.

31: Logics of Programs. Proceedings, 1981. Edited by D. Kozen. 9 pages. 1982.

32: Data Base Design Techniques I: Requirements and Logical tures. Proceedings, 1978. Edited by S.B. Yao, S.B. Navathe, Veldon, and T.L. Kunii. V, 227 pages. 1982.

33: Data Base Design Techniques II: Proceedings, 1979. d by S.B. Yao and T.L. Kunii. V, 229–399 pages. 1982.

34: Program Specification. Proceedings, 1981. Edited by J. nstrup. IV, 426 pages. 1982.

35: R.L. Constable, S.D. Johnson, and C.D. Eichenlaub, An duction to the PL/CV2 Programming Logic. X, 292 pages. 1982.

36: Ch. M. Hoffmann, Group-Theoretic Algorithms and Graph orphism. VIII, 311 pages. 1982.

37: International Symposium on Programming. Proceedings, Edited by M. Dezani-Ciancaglini and M. Montanari. VI, 406 s. 1982.

38: 6th Conference on Automated Deduction. Proceedings, Edited by D.W. Loveland. VII, 389 pages. 1982.

39: J. Uhl, S. Drossopoulou, G. Persch, G. Goos, M. Dausmann, Vinterstein, W. Kirchgässner, An Attribute Grammar for the antic Analysis of Ada. IX, 511 pages. 1982.

40: Automata, Languages and programming. Edited by M. Niel- and E.M. Schmidt. VII, 614 pages. 1982.

41: U. Kastens, B. Hutt, E. Zimmermann, GAG: A Practical piler Generator. IV, 156 pages. 1982.

Vol. 142: [...] ftware Productio[...] Murli. VII, 271 p[...]

Vol. 143: [...] Edited by M. Ma[...]

Vol. 144: [...] Calmet. XIV, 301 pa[...]

Vol. 145: Theoretical Computer Science. Proceedings, 1983. Edited by A.B. Cremers and H.P. Kriegel. X, 367 pages. 1982.

Vol. 146: Research and Development in Information Retrieval. Proceedings, 1982. Edited by G. Salton and H.-J. Schneider. IX, 311 pages. 1983.

Vol. 147: RIMS Symposia on Software Science and Engineering. Proceedings, 1982. Edited by E. Goto, I. Nakata, K. Furukawa, R. Nakajima, and A. Yonezawa. V. 232 pages. 1983.

Vol. 148: Logics of Programs and Their Applications. Proceedings, 1980. Edited by A. Salwicki. VI, 324 pages. 1983.

Vol. 149: Cryptography. Proceedings, 1982. Edited by T. Beth. VIII, 402 pages. 1983.

Vol. 150: Enduser Systems and Their Human Factors. Proceedings, 1983. Edited by A. Blaser and M. Zoeppritz. III, 138 pages. 1983.

Vol. 151: R. Piloty, M. Barbacci, D. Borrione, D. Dietmeyer, F. Hill, and P. Skelly, CONLAN Report. XII, 174 pages. 1983.

Vol. 152: Specification and Design of Software Systems. Proceedings, 1982. Edited by E. Knuth and E. J. Neuhold. V, 152 pages. 1983.

Vol. 153: Graph-Grammars and Their Application to Computer Science. Proceedings, 1982. Edited by H. Ehrig, M. Nagl, and G. Rozenberg. VII, 452 pages. 1983.

Vol. 154: Automata, Languages and Programming. Proceedings, 1983. Edited by J. Díaz. VIII, 734 pages. 1983.

Vol. 155: The Programming Language Ada. Reference Manual. Approved 17 February 1983. American National Standards Institute, Inc. ANSI/MIL-STD-1815A-1983. IX, 331 pages. 1983.

Vol. 156: M.H. Overmars, The Design of Dynamic Data Structures. VII, 181 pages. 1983.

Vol. 157: O. Østerby, Z. Zlatev, Direct Methods for Sparse Matrices. VIII, 127 pages. 1983.

Vol. 158: Foundations of Computation Theory. Proceedings, 1983. Edited by M. Karpinski, XI, 517 pages. 1983.

Vol. 159: CAAP'83. Proceedings, 1983. Edited by G. Ausiello and M. Protasi. VI, 416 pages. 1983.

Vol. 160: The IOTA Programming System. Edited by R. Nakajima and T. Yuasa. VII, 217 pages. 1983.

Vol. 161: DIANA, An Intermediate Language for Ada. Edited by G. Goos, W.A. Wulf, A. Evans, Jr. and K.J. Butler. VII, 201 pages. 1983.

Vol. 162: Computer Algebra. Proceedings, 1983. Edited by J.A. van Hulzen. XIII, 305 pages. 1983.

Vol. 163: VLSI Engineering. Proceedings. Edited by T.L. Kunii. VIII, 308 pages. 1984.

Vol. 164: Logics of Programs. Proceedings, 1983. Edited by E. Clarke and D. Kozen. VI, 528 pages. 1984.

Vol. 165: T.F. Coleman, Large Sparse Numerical Optimization. V, 105 pages. 1984.

Vol. 166: STACS 84. Symposium of Theoretical Aspects of Computer Science. Proceedings, 1984 Edited by M. Fontet and K. Mehlhorn. VI, 338 pages. 1984.

Lecture Notes in Computer Science

Edited by G. Goos and J. Hartmanis

211

Uwe Schöning

Complexity and Structure

Springer-Verlag
Berlin Heidelberg New York Tokyo

Author

Uwe Schöning
Erziehungswissenschaftliche Hochschule Rheinland-Pfalz
Seminar für Informatik
Rheinau 3–4, D-5400 Koblenz,

CR Subject Classifications (1985): F.1, F.4.1, G.3

ISBN 3-540-16079-5 Springer-Verlag Berlin Heidelberg New York Tokyo
ISBN 0-387-16079-5 Springer-Verlag New York Heidelberg Berlin Tokyo

Printing and binding: Beltz Offsetdruck, Hemsbach/Bergstr.
2145/3140-543210

PREFACE

This monograph is a revised and extended version of a thesis submitted to the faculty of Mathematics and Computer Science at the University of Stuttgart, Germany, in partial fulfillment of the German "Habilitation" degree. It summarizes certain aspects in structural complexity theory related to the author´s research and interests over the last years.

Many people contributed to this work. It is a pleasure to thank Prof. W. Schwabhäuser and Prof. R. Book for their valuable advice, not only in scientific questions. I am most grateful to the Deutsche Forschungsgemeinschaft for the financial support during the last years. I benefitted highly from many discussions and ideas that have been shared with me by José Balcázar, Ker-I Ko, Pekka Orponen and David Russo.

Nina deserves special thanks for her love and patience.

Stuttgart, September 1985 U.S.

CONTENTS

CONTENTS

INTRODUCTION

Complexity theory is dealing with the classification of the recursive sets and functions. Mostly, these classifications are quantitative, in the sense that the amount of computational resources (i.e. computation time and space) is investigated that is sufficient (upper bounds) and that is neccessary (lower bounds) to compute certain functions, or to decide certain problems. The aim of this "quantitative" complexity theory is to prove tight upper and lower bounds for as big classes of functions (or problems) as possible. The ideal case is that the upper and lower bounds coincide.

In this monograph we discuss certain aspects of a complexity theory that might be called "qualitative" or "structural". Structural complexity is strongly influenced by recursion theory. Many notions and ideas stemming from recursion theory have been successfully transformed into the context of complexity theory (notions like "p-selective", "immune", or "low" and "high"). On the other hand, some notions do not seem to be transformable, or do not seem to make sense in complexity theory, like the "relativization principle". But one aim in structural complexity theory is to give explanations for such phenomena.

An important task in structural complexity theory is to show logical implications or equivalences between certain unresolved problems, and hence, to give a certain insight in the nature of such problems. The most outstanding open problem in this context is the P-NP-problem, i.e. the question whether nondeterminism really "helps" in polynomial time bounded computations. The theory of NP-completeness has achieved much attention since the fundamental papers by Cook (1971) and Karp (1972). Hundreds of problems from various areas in Mathematics and Computer Science, such as Graph Theory, Game Theory, Number Theory, Operations Research, Logic, Combinatorics, etc. are now known to be NP-complete (cf. Garey and Johnson, 1979).

In our understanding, the notion of "NP-completeness" is a typical structural one. It has not much to do with the complexity hierarchies in quantitative complexity theory: although NP-complete sets can be understood as the hardest problems in the class NP, most of them can be solved in nondeterministic (quasi-)linear time, i.e. they are very low in the standard nondeterministic time hierarchy. The importance of the NP-completeness notion is due to the following logical connection: NP-complete sets are solvable in polynomial time if and only if P = NP.

Summarizing we can say, structural complexity theory is concerned with explanations for the hardness of certain problems (e.g. NP-complete problems). A typical "explanatory" notion is the "complexity core" of a problem, that is a collection of instances which is "uniformly hard", independent of the particular algorithm used. Here, the question is addressed whether intractable problems are hard because of the existence of many particular hard instances, or whether they are hard because of the particular "conglomeration" of instances.

Motivated by the still unsolved P-NP-question, many structural methods have been developed, and even "classical" concepts have been "rediscovered" to be useful in the context of polynomial time computations. One of these classical concepts is Boolean circuit complexity which will play a central role in this work. Circuits are purely combinatorial objects, and their behavior is intuitively well understood. Therefore, the comparison of the "clean" circuit complexity with the usual complexity that depends on the particular machine model used (Turing machines, RAMs), is an interesting topic. Furthermore, when circuits are used to describe languages (hence infinite objects), the obtained complexity measure is called nonuniform, as opposed to the uniform measure obtained by, say, Turing machine complexity. At the borderline between uniform and nonuniform models, there is a variety of models and complexity classes, based on probabilistic or on approximation algorithms, that can be investigated, especially when applied to NP-complete sets.

The notion of polynomially bounded circuit-size complexity, later denoted P/Poly, will be the red thread in this monograph. It will be shown that many concepts, definitions, results can be much easier described and understood in terms of this notion. The contents of this monograph have been placed together using this guideline and the author's preferences. It is not intended to be complete.

We introduce some basic notions and definitions in chapter 1. The reader should be familiar with the basics of complexity theory, as covered by a textbook like (Hopcroft and Ullman, 1979). Chapter 2 introduces the central notion of Boolean circuit-size complexity. Some classical results of circuit complexity are presented as far as they are relevant for the sequel. A circuit model is presented that compares to the "standard" model like NP to P. Thus, in a sense, a projection of the P-NP-problem into the context of circuit complexity is given. The central notion of "polynomial size circuits" will be introduced in section 2.4, and the "p-selective" sets will be shown to possess polynomial size circuits in section 2.5.

The theme of the third chapter is the classification of probabilistic algorithms and the complexity classes obtained by the respective probabilistic algorithms. It is shown that certain sets recognizable by probabilistic Turing machines have polynomial circuit-size complexity. In fact, the class BPP (bounded error probabilistic polynomial time) can be characterized very similar to the definition of "polynomial size circuits". This requires to analyze how the behavior of a probabilistic algorithm improves if it is "iterated" several times.

"Sparse" sets are studied in chapter 4. The study of density notions in complexity theory was initiated by the work of Berman and Hartmanis (1977) who studied the question whether all NP-complete sets are polynomially isomorphic (i.e. the "Berman-Hartmanis-conjecture"). The question whether sparse sets in NP - P, or in particular, sparse NP-complete sets can exist, is discussed in section 4.1. Sparse sets can be used to define certain approximations to intractable sets. Two such approaches are studied in sections 4.3 and 4.4, and the results obtained state that intractable problems (like NP-complete problems) are not even efficiently solvable in such an approximate sense (unless P = NP). In this context the notion of a complexity core is introduced, and a main result states that NP-complete sets have non-sparse complexity cores. Finally, sparse sets (and sets over one-letter alphabets) are used to give a characterization of the class P/Poly (i.e. the sets having polynomial size circuits), and in section 4.5, a theorem about "self-reducible" sets is proved that will have several applications in the next chapters.

Chapter 5 introduces "low" and "high" hierarchies of languages (in particular, within NP). Low sets behave similar to P-sets, they have low "information content", whereas high sets behave similar to NP-complete sets, in fact, highness can be understood as generalized NP-completeness. No set can be both low and high unless some unlikely event happens (the polynomial hierarchy "collapses"). A somewhat surprising link to circuit complexity will be drawn in section 5.2. Polynomial circuit-size complexity is shown to be a lowness property. The lowness of some specific classes, like BPP, is studied in sections 5.3 - 5.5.

Some of the methods and results developed so far have interesting consequences in the relativized complexity theory. A well known phenomenon in complexity theory is that the "relativization principle" from recursive function theory does not hold. E.g., there are oracles A and B such that P(A) = NP(A), but P(B) ≠ NP(B) (Baker, Gill and Solovay, 1975). Certain explanations for this phenomenon are given in terms of "positive" and "negative" relativizations (sections 6.1 and 6.2). The behavior of oracles in higher levels of the polynomial hierarchy is studied in section 6.3.

CHAPTER 1

PRELIMINARIES

In the following we introduce some of the notations and definitions which will be needed throughout this monograph.

All the sets in this work will be languages over the fixed "standard" alphabet $\Sigma = \{0,1\}$. The empty string is denoted by λ. For a string $w \in \Sigma^*$ let $|w|$ be its length, and for a set A let $|A|$ be its cardinality. We assume the standard total ordering $<$ on Σ^* such that smaller strings preceed longer ones, and within the same length strings are ordered lexicograhically.

Define the complement of a set A, $\overline{A}$, as follows: first, let Γ be the smallest alphabet such that $A \subseteq \Gamma^*$, then let $\overline{A} = \Gamma^* - A$. (The use of this definition is that complements of tally sets - sets over a one letter alphabet - become tally, too).

Let N be the set of natural numbers (including zero). For a set $A \subseteq \Sigma^*$ and an $n \in N$ define

$$A_{=n} = \{ x \in A \mid |x| = n \},$$

and similarly,

$$A_{\leq n} = \{ x \in A \mid |x| \leq n \}.$$

The symmetric difference of two sets A and B is $A \Delta B = (A - B) \cup (B - A)$, and the "marked union" is

$$A \oplus B = \{ 0x \mid x \in A \} \cup \{ 1x \mid x \in B \}.$$

The indicator function of a set A is the function $\chi_A : \Sigma^* \to \{0,1\}$,

$$\chi_A(x) = \begin{cases} 1, & x \in A \\ 0, & \text{otherwise.} \end{cases}$$

For a real number x, let $[x]$ be the largest integer n such that $n \leq x$. We use log for the logarithm of base 2. In the following a polynomial is understood as a function p of the form

$$p(n) = a_k n^k + \dots + a_1 n + a_0, \quad a_i \in N, \; a_k = 0, \quad 0 \leq i \leq k.$$

Thus, polynomials are always nondecreasing. The degree of this polynomial p is k.

For a function f, $O(f)$ denotes a function g with the property that there is a constant $c > 0$ such that $g(n) \leq c \cdot f(n)$ for almost every n. Conversely, $\Omega(f)$ denotes a function g such that f is a function of the type $O(g)$.

Our model of computation is the (multi-tape) Turing machine. For formal definitions see the textbook of Hopcroft and Ullman (1979). We will consider several types of Turing machines. A Turing machine acceptor has a distinguished accepting (final) state. The language language accepted by (deterministic or nondeterministic) Turing machine M is denoted $L(M)$.

A Turing machine can also be used to compute a function from Σ^* to Σ^*. Such a transducer has a distinguished output tape. In general, transducers are deterministic.

An oracle Turing machine has a seperate oracle tape and three distinguished states: the query state, the yes-state, and the no-state. During the course of a computation the oracle Turing machine can write some string w on the output tape, and after entering the query state it transfers "magically" into the yes-state or no-state depending on whether the string on the oracle tape is in the "oracle language" or not. After such an oracle query the oracle tape is instantly erased. For an oracle Turing machine M (deterministic or nondeterministic) and set A, let $L(M, A)$ be the set accepted by M when using A as its oracle language.

A Turing machine M is $t(n)$ time bounded for some function t on the natural numbers if each computation of M on inputs of size n has length at most $t(n)$. M is $t(n)$ space bounded if no computation of M on inputs of size n uses more than $t(n)$ tape squares. If t can be chosen to be a polynomial then M is polynomially time (or space, resp.) bounded. Define the following classes of languages:

DTIME(t(n)) = { L(M) | M is deterministic and t(n) time bounded },

NTIME(t(n)) = { L(M) | M is nondeterministic and t(n) time bounded },

DSPACE(s(n)) = { L(M) | M is deterministic and s(n) space bounded },

NSPACE(s(n)) = { L(M) | M is nondeterministic and s(n) space bounded}.

For results comparing these classes see any textbook on complexity theory (e.g. Hopcroft and Ullman, 1979). Of particular interest for us are the following classes defined in terms of polynomially (and exponentially) time or space bounded Turing machines.

$P = \bigcup \{DTIME(p(n)) \mid p \text{ is a polynomial}\}$,

$NP = \bigcup \{NTIME(p(n)) \mid p \text{ is a polynomial}\}$,

$PSPACE = \bigcup \{DSPACE(p(n)) \mid p \text{ is a polynomial}\}$,

$EXPTIME = \bigcup \{DTIME(2^{cn}) \mid c > 0\}$,

$NEXPTIME = \bigcup \{NTIME(2^{cn}) \mid c > 0\}$,

$EXPSPACE = \bigcup \{DSPACE(2^{cn}) \mid c > 0\}$.

Note that by "Savitch's Theorem" (cf. Hopcroft and Ullman, 1979) it makes no difference whether we take deterministic or nondeterministic machines in the definitions of PSPACE and EXPSPACE.

The following inclusions are easy to see, but none of them is known to be proper.

$P \subseteq NP \subseteq PSPACE$, and

$EXPTIME \subseteq NEXPTIME \subseteq EXPSPACE$.

From the time and space hierarchy theorems (cf. Hopcroft and Ullman, 1979) it follows that

$P \subsetneq EXPTIME$, $NP \subsetneq NEXPTIME$, and $PSPACE \subsetneq EXPSPACE$.

The most famous open problem within this context is the well known "P-NP-problem", i.e. the question wether $P = NP$ or $P \neq NP$ holds. Also unknown are the answers to the following questions:

$P = PSPACE$?

$NP = PSPACE$?

EXPTIME = NEXPTIME ?

EXPTIME = EXPSPACE ?

NEXPTIME = EXPSPACE ?

For each class of languages C, let co-C be the set of complements of the sets in C, co-C = { A | $\overline{A} \in$ C }. Classes that are defined via deterministic time or space bounded Turing machines (i.e. P, EXPTIME, and EXPSPACE) are closed under complementation. For the nondeterministic classes, these are open questions:

NP = co-NP ?

NEXPTIME = co-NEXPTIME ?

For each of the above classes C and set A, the relativized class C(A) can be defined taking oracle Turing machines instead of usual Turing machines in the respective definitions (i.e. taking L(M, A) instead of L(M).) Obviously, C = C(∅). Furthermore, for a class of sets D, define C(D) = $\bigcup$ { C(A) | A $\in$ D }.

We assume some standard way of encoding combinatorial objects such as Boolean formulas, circuits, graphs etc. into strings over Σ. Modulo polynomial time computations, the particular way of encoding does not make any difference. The only care that has to be taken is in encoding integers, they should be represented in binary - not in unary. Though sometimes, for certain padding reasons, it is convenient to encode integers in unary. But then this will be explicitely stated.

Further, we assume there is a way of pairing strings. Let $\langle x, y\rangle \in \Sigma^*$ denote the encoding of the strings $x \in \Sigma^*$ and $y \in \Sigma^*$. The pairing function and its inverses should be computable in polynomial time. Such a pairing function can be extended in a standard fashion to encode arbitrary finite sequences of strings.

If x and y are tally strings (over a one-letter alphabet) then also $\langle x, y\rangle$ is assumed to be tally.

Definition 1.1. (Stockmeyer, 1977) The polynomial-time hierarchy PH consists of the classes Σ_k^p, Π_k^p, and Δ_k^p for k=0,1,2,... . These are defined inductively as follows.

$\Sigma_0^P = \Pi_0^P = \Delta_0^P = P$, and for each $k \geq 0$,
$\Sigma_{k+1}^P = NP(\Sigma_k^P)$,
$\Pi_{k+1}^P = \text{co-}NP(\Sigma_k^P) = \text{co-}\Sigma_{k+1}^P$,
$\Delta_{k+1}^P = P(\Sigma_k^P)$.
Further, let $PH = \bigcup\{ \Sigma_k^P \mid k \geq 0 \}$.

From the definition it follows immediately that $\Delta_1^P = P$, $\Sigma_1^P = NP$, and $\Pi_1^P = \text{co-}NP$. Further, $PH \subseteq PSPACE$. For each $k \geq 1$ the following inclusions hold.

$$\Delta_k^P \subseteq \Sigma_k^P \cap \Pi_k^P \begin{array}{c} \subseteq \Sigma_k^P \subseteq \\ \\ \subseteq \Pi_k^P \subseteq \end{array} \Sigma_k^P \cup \Pi_k^P \subseteq \Delta_{k+1}^P .$$

There is another characterization of the polynomial hierarchy in terms of polynomially bounded quantifiers (cf. Stockmeyer, 1977; Wrathall 1977). It will be used very frequently.

$A \in \Sigma_k^P$ if and only if there exist polynomials $p_1,\ldots,p_k$ and a set $B \in P$ such that

$$A = \{ x \mid (\exists y_1, |y_1| \leq p_1(|x|))(\forall y_2, |y_2| \leq p_2(|x|)) \ldots (Q_k\, y_k, |y_k| \leq p_k(|x|)) \; \langle x,y_1,\ldots,y_k\rangle \in B \} .$$

Here, $Q_k = \exists$ if k is odd, and $Q_k = \forall$ if k is even. A similar dual characterization of Π_k^P can be given where the $\forall$-quantifier is used first. In the following we will use the abbreviations

$$(\exists y)_p \quad \text{and} \quad (\forall y)_p$$

for

$$(\exists y, |y| \leq p(|x|)) \quad \text{and} \quad (\forall y, |y| \leq p(|x|)), \quad \text{resp.}$$

Since finitely many polynomials can always be majorized by another polynomial, and since polynomials are time-constructible (see Hopcroft and Ullman, 1979), the polynomials $p_1,\ldots,p_k$ above can be chosen identical. Further, in some applications it is more convenient to use the following characterization: $A \in \Sigma_k^P$ if and only if there is a polynomial p and a set $B \in P$ such that

$$A = \{ x \mid (\exists y_1, |y_1| = p(|x|)) (\forall y_2, |y_2| = p(|x|)) \ldots (Q_k\, y_k, |y_k| = p(|x|)) \; \langle x, y_1, \ldots, y_k \rangle \in B \}.$$

It is not known whether the polynomial-time hierarchy consists of infinitely many distinct classes (i.e. $\Sigma_k^P \neq \Sigma_{k+1}^P$ for each k) or whether it "collapses" (i.e. for some k, $\Sigma_k^P = \Sigma_{k+1}^P = \ldots = PH$). The first case implies $PH \neq PSPACE$ (Wrathall, 1977).

We will consider two types of polynomial time bounded reducibilities. A set $A \subseteq \Sigma^*$ is polynomial time <u>many-one reducible</u> to a set $B \subseteq \Sigma^*$ (in symbols: $A \leq_m^P B$) if there is a function $f : \Sigma^* \to \Sigma^*$ computable by a polynomial time bounded Turing machine transducer such that for each $x \in \Sigma^*$, $x \in A$ iff $f(x) \in B$ (cf. Karp, 1972).

Further, A is polynomial time <u>Turing reducible</u> to B (in symbols: $A \leq_T^P B$) if $A \in P(B)$ (cf. Cook, 1971).

Observe that for two sets A and B, $A \oplus B$ is the least upper bound with respect to the partial order $\leq_m^P$ (and also $\leq_T^P$).

Let r be a relation on languages (e.g., $r = \leq_m^P$ or $r = \leq_T^P$). A set A is called r-<u>hard</u> for a class of sets C if for each $B \in C$, $B \; r \; A$. A is called r-<u>complete</u> for C if, additionally, $A \in C$. If the type of reducibility is understood (mostly $\leq_m^P$) or not important, then A is sometimes just called "C-hard" or "C-complete". The C-complete languages form, in a sense, the "most difficult" sets within the class C. Especially interesting is the case that $C = NP$. Let A be an NP-complete set. Then,

$$P = NP \text{ if and only if } A \in P.$$

The above line gives the reason for the importance of the theory of NP-completeness and the P-NP-question. There are hundreds of languages (encodings of combinatorial problems) now known to be NP-complete (cf. Garey and Johnson, 1979). These problems stem from all kinds of areas within Mathematics and Computer Science, and many researchers independently tried to find efficient (i.e. polynomial time bounded) algorithms for them, but didn't succeed. This is generally taken as evidence that no such algorithms exist, and thus, by the above line, $P \neq NP$. A formal proof of this still seems to be beyond the present capabilities.

We need to fix some particular NP-complete language which will be used later several times. We choose the satisfiability problem which was the first problem shown to be NP-complete (Cook, 1971).

$$SAT = \{ \langle F \rangle \mid F \text{ is a satisfiable Boolean formula } \}.$$

Further, we fix a language known to be PSPACE-complete (Stockmeyer, 1977).

$$QBF = \{ \langle F \rangle \mid F \text{ is a quantified Boolean formula in prefix form without free variables which is true } \}.$$

For more details about quantified Boolean formulas see (Hopcroft and Ullman, 1979, pp. 343). It is shown in (Wrathall, 1977; Stockmeyer, 1977) that for each $k \geq 1$ the following language QBF_k is Σ_k^P-complete.

$$QBF_k = \{ \langle F \rangle \mid \langle F \rangle \in QBF \text{ and the quantifier prefix of } F \text{ starts with "}\exists\text{", and there are at most } k-1 \text{ alternations of quantifiers } \}.$$

The following facts about the polynomial-time hierarchy will be used several times: If A is a Σ_k^P-complete set, then $\Sigma_n^P(A) = \Sigma_{n+k}^P$, $n,k \geq 1$. Further, for all $n,k \geq 1$,

$$\Sigma_n^P(\Sigma_k^P \cap \Pi_k^P) = \Sigma_{n+k-1}^P,$$

i.e., in particular, we have $NP(NP \cap co\text{-}NP) = NP$.

CHAPTER 2

CIRCUIT-SIZE COMPLEXITY

In this chapter we develop some foundations of circuit-size complexity which is needed for the following investigations. One of our aims is to show that the concept of circuit-size complexity is also useful in other areas such as probabilistic algorithms, approximation algorithms for intractable problems, relativizations, and others.

The circuit-size complexity of a Boolean function is the smallest number of gates in a Boolean circuit that computes the function. In chapter 2.1 we present the classical results by Lupanov (1958) and Shannon (1949) which assert that, in a sense, most n-ary Boolean functions have circuit-size complexity of order $2^n/n$. Chapter 2.2 introduces the concept of circuit-size complexity applied to languages (hence infinite objects), and a central notion for this monograph is introduced: polynomial size circuit complexity. Later this class will be formally denoted by P/Poly, a notation introduced by Karp and Lipton (1980).

A new circuit model is introduced in chapter 2.3 which, in a sense, "generates" (instead "accepts") the language in question. It is shown that this model is related to the original model like the complexity class NP to P. In fact, strong relationships to the P-NP-question are drawn.

Finally, an interesting class of sets, the p-selective languages, is shown to have polynomial circuit-size complexity.

2.1. CIRCUIT-SIZE COMPLEXITY OF BOOLEAN FUNCTIONS

In the following we will restrict ourselves to Boolean circuits having gates with exactly two inputs where each gate can compute one of the 16 different 2-ary Boolean functions. It is a trivial fact that each Boolean function can be computed by a Boolean circuit of this type. We measure the circuit-size complexity of a Boolean function as the minimum number of gates necessary to compute the function. Formally we introduce circuits as sequences of "gates" where each input of a gate is either connected to an output of a previous gate or it forms a general input to the circuit.

<u>Definition 2.1.</u> A <u>circuit with n inputs</u> is a finite sequence $c = (g_{n+1}, \ldots, g_{n+q})$, $q \geq 1$, such that each g_r, $n+1 \leq r \leq n+q$, is a triple (i_r, j_r, k_r) with $i_r \in \{1,\ldots,16\}$ and $1 \leq j_r, k_r < r$. Each such triple is called a <u>gate</u>. The <u>size</u> of this circuit is q. If $f_1,\ldots,f_{16}$ is a fixed enumeration of all 16 2-ary Boolean functions, then we define the <u>r-th function computed by c</u>, $res_c^r : \{0,1\}^n \to \{0,1\}$, $1 \leq r \leq n+q$, as follows:

$$res_c^r(x_1,\ldots,x_n) = x_r \quad \text{for } 1 \leq r \leq n,$$
$$res_c^r(x_1,\ldots,x_n) = f_{i_r}(res_c^{j_r}(x_1,\ldots,x_n), res_c^{k_r}(x_1,\ldots,x_n))$$
$$\text{for } n+1 \leq r \leq n+q.$$

The (final) <u>function computed by c</u>, denoted f_c, is res_c^{n+q}. The <u>circuit-size complexity</u> of a Boolean function f is defined as the natural number

$$C(f) = \min \{ q \mid \text{there is a circuit of size } q \text{ that computes } f \}.$$

The diagram on the next page illustrates the definition.

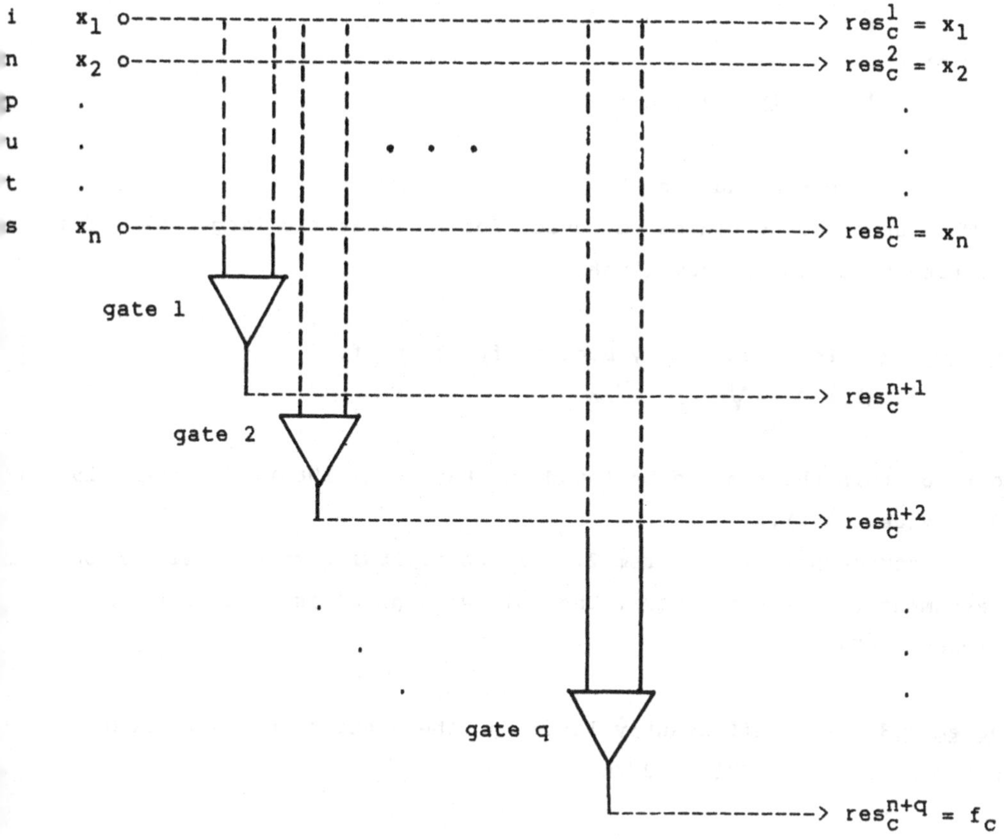

Now we want to find tight upper and lower bounds for the circuit-size complexity of arbitrary Boolean functions. It is easy to see that for each n-ary Boolean function f, $C(f) \leq 2^n - 3$ $(n \geq 2)$. This follows from the observation that each n-ary Boolean function can be written in the following way:

$$f(x_1,\dots,x_n) = (x_1 \wedge f(1,x_2,\dots,x_n)) \vee (\neg x_1 \wedge f(0,x_2,\dots,x_n))$$

From this representation of f we obtain

$$C(f) \leq g(n) \quad (n \geq 2)$$

where g is uniquely given by the scheme

$$g(n) = \begin{cases} 1 & , n = 2 \\ 2g(n-1) + 3 & , n > 2. \end{cases}$$

It is easy to check that $g(n) = 2^n - 3$.

This upper bound can be improved (for sufficiently large n). This is a result due to Lupanov (1958).

<u>Theorem 2.2</u>. For each n-ary Boolean function f,

$$C(f) \leq (1 + O(1/\sqrt{n})) \cdot 2^n/n.$$

Proofs of this theorem can be found in (Harrison, 1965), (Fischer, 1974), and (Savage, 1976).

The counterpart to Theorem 2.2 is the following result going back to an argument of Shannon (1949). The following proof is adapted from (Fischer, 1974).

<u>Theorem 2.3</u>. For sufficiently large n there exist n-ary Boolean functions f with $C(f) > 2^n/n$.

<u>Proof</u>. The proof proceeds by a counting argument. We compare the total number of different n-ary Boolean functions with an upper bound to $N_{q,n}$ which is the number of different n-ary Boolean functions that can be computed by circuits of size at most q. By this comparison we obtain a lower bound for q.

There exist $M = [16(n+q)^2]^q$ different sequences c of the form $c = (g_{n+1},\ldots,g_{n+q})$ such that $g_r = (i_r,j_r,k_r)$, $n+1 \leq r \leq n+q$, $1 \leq i_r \leq 16$, $1 \leq j_r,k_r \leq n+q$. Among these sequences are all circuits of size q. But many of these sequences are not permissible circuits (because $j_r \geq r$ or $k_r \geq r$), and among the circuits many compute the same functions. The following is to take this into account.

A circuit c is called <u>reduced</u>, if for all r,s with $n+1 \leq r < s \leq n+q$ holds $res_c^r \neq res_c^s$. Let $\pi : \{1,\ldots,q\} \to \{1,\ldots,q\}$ be a permutation and let c be a sequence as above. Then we define $\pi(c)$ as

$$\pi(c) = (g'_{n+1},\ldots,g'_{n+q}) \quad \text{where} \quad g'_{n+\pi(r)} = (i_r,j'_r,k'_r),$$

$$j'_r = \begin{cases} j_r & , \quad j_r \le n \\ n+\pi(d_r) & , \quad j_r = n+d_r \end{cases},$$

$$k'_r = \begin{cases} k_r & , \quad k_r \le n \\ n+\pi(d_r) & , \quad k_r = n+d_r \end{cases}.$$

The idea of this definition is this: if both c and $\pi(c)$ are circuits according to Definition 2.1, then gate $g'_{n+\pi(r)}$ in $\pi(c)$ does the same job as $g_{n+\pi(r)}$ in c. The corresponding connections from the previous gates are described by j'_r and k'_r. Hence, if both c and $\pi(c)$ are circuits then $\{res_1^c,\ldots,res_q^c\} = \{res_1^{\pi(c)},\ldots,res_q^{\pi(c)}\}$, i.e. c and $\pi(c)$ compute the same set of Boolean functions.

Further, if c is a reduced circuit of size q, then all $q!$ sequences of the form $\pi(c)$ where π is a permutation of $\{1,\ldots,q\}$ are pairwise different. It follows that for each reduced circuit of size q there are at least $q!$ different sequences among the above M sequences which compute the same set of q functions as c or which are no permissible circuits.

The next observation is that for each n-ary Boolean function f with $C(f) \le q$ there is a reduced circuit of <u>exactly</u> size q that computes f. (First choose a smallest - hence reduced - circuit d that computes f, then fill d up systematically with additional gates that compute functions not yet computed in d. This argument goes through if q does not exceed the number of all n-ary Boolean functions, i.e. $q \le 2^{2^n}$).

Now we can conclude:

$$N_{q,n} \le (Mq)/q! = [16(n+q)^2]^q \; / \; (q-1)!$$

From Stirling's approximation of the factorial function we have

$$q! \ge \sqrt{2\pi q} \cdot (q/e)^q .$$

Hence we get

$$N_{q,n} \leq \frac{q}{\sqrt{2\pi q}} \cdot \left(\frac{16e(n+q)^2}{q}\right)^q < q(64eq)^q \quad \text{for} \quad q \geq n.$$

Thus, $N_{q,n} < q(cq)^q$ where $c = 64e = 173.97\ldots$. Since there are 2^{2^n} different n-ary Boolean functions, the circuit size $q = q(n)$ necessary to be able to compute all of them must satisfy the inequality

$$N_{q,n} \geq 2^{2^n} .$$

Letting $q = 2^n/n$, we get

$$\begin{aligned} N_{2^n/n,n} &< (2^n/n) \cdot (c \cdot 2^n/n)^{2^n/n} \\ &< (2^n/n) \cdot 2^{-2^n/n} \cdot 2^{2^n} \quad \text{for} \quad n > 2c \\ &< 2^{2^n} \quad \text{because} \quad (2^n/n)\, 2^{-2^n/n} < 1. \end{aligned}$$

Hence it follows that (for sufficiently large n) $q = 2^n/n$ gates do not suffice to compute all n-ary Boolean functions. Thus there exist n-ary Boolean functions f with $C(f) > 2^n/n$. □

A refinement of the above argument shows that the ratio of n-ary Boolean functions f with $C(f) < (1-e)\, 2^n/n$ is smaller than 2^{-e2^n}, i.e. goes exponentially to 0 as e -> 1. Hence, in a sense, "most" n-ary Boolean functions have circuit-size complexity of order $2^n/n$.

Interestingly, the counting argument in the proof of Theorem 2.3 does not give any hint how a "natural" class of Boolean functions having circuit-size complexity $2^n/n$ might look like. Not even non-linear lower bounds for such "natural" classes are known.

An "unnatural" - but well defined - class of Boolean functions $\{f_n\}$ of circuit-size complexity at least $2^n/n$ would be the following:

f_n = the first n-ary Boolean function f (according to some canonical ordering of the Boolean functions) having $C(f) > 2^n/n$.

2.2. CIRCUIT-SIZE COMPLEXITY OF LANGUAGES

Now we extend the notion of circuit-size complexity from Boolean functions (finite objects) to languages over $\{0,1\}$ (infinite objects). The complexity will now become a function on the natural numbers.

Let $A \subseteq \{0,1\}^*$ be a language. For each $n \geq 0$, let χ_A^n : $\{0,1\}^n \to \{0,1\}$,

$$\chi_A^n(x) = \begin{cases} 1 , & x \in A \\ 0 , & \text{otherwise} \end{cases}$$

be the indicator function of A for strings of size n.

Next we define the most important notion for all to follow.

Definition 2.4. The circuit-size complexity of a language $A \subseteq \{0,1\}^*$ is the function $CS_A(n) = C(\chi_A^n)$. A set A has polynomially bounded circuit-size complexity (or, polynomial size circuits) if there is a polynomial p such that for each $n \geq 0$, $CS_A(n) \leq p(n)$.

The circuit-size complexity of a set is sometimes refered to as the "non-uniform" complexity of the set - as opposed to the "uniform" (deterministic Turing machine) complexity. It is called non-uniform since it is defined in terms of a, in general, non-uniform sequence of computational devices (here circuits) as opposed to one uniform Turing machine accepting the whole set. In general, the function mapping n to the n-th circuit that computes χ_A^n need not be recursive. We will discuss the relationship between the complexity of the set and the above mappings in a later chapter.

Each language over the one-letter alphabet $\{0\}$ (i.e. "tally" languages) has polynomially bounded (even linearly bounded) circuit-size complexity. On the other hand, by simple diagonalization, there are arbitrarily complex tally languages (i.e. for each recursive function t there is a tally language not in DTIME(t(n))). Hence there exist arbitrarily complex languages having polynomial size circuits. An upper bound for the circuit-size complexity of a set does not necessarily imply any upper bound for the (Turing machine) complexity of that set (but we will prove other implications later).

On the other hand, we will later show that an upper bound for the (deterministic Turing machine) complexity of a set does imply a certain upper bound for its circuit-size complexity.

From Theorem 2.3 it follows directly that there exist sets not having polynomial size circuits. But the theorem does not make an assertion about the recursiveness of such sets. Next we show that such recursive sets can be found. The proof is a modification of an argument found in (Kannan, 1982).

Theorem 2.5. There exist sets in EXPSPACE not having polynomial size circuits.

Proof. We define such a set in stages. In stage n the sets $\emptyset = A_{n,0} \subseteq A_{n,1} \subseteq \cdots \subseteq A_{n,2^n} \subseteq \{0,1\}^n$ are constructed where $A_{n,i} - A_{n,i-1}$ is either the empty set or $\{x_i^n\}$. Here x_i^n is the i-th string of size n (according to lexicographical order), $1 \leq i \leq 2^n$. Finally, we define A as $\bigcup_n A_{n,2^n}$.

```
stage n:
  A_{n,0} := Ø ;
  for i := 1 to 2^n do
    begin
      zeros := 0 ;
      num := 0 ;
      for each w ∈ {0,1}*, |w| < n^(log n), such that w is encoding of
      a circuit c with n inputs do
        if c computes A_{n,i-1} correctly on inputs x_1^n,...,x_{i-1}^n then
          begin
            num := num + 1 ;
            if f_c(x_i^n)=0 then zeros := zeros + 1 ;
          end;
      if zeros > num/2 then A_{n,i} := A_{n,i-1} U {x_i^n}
                       else A_{n,i} := A_{n,i-1} ;
    end;
end of stage n.
```

Suppose, A has polynomial size circuits. Then for a polynomial p and each $n \geq 0$ there is a circuit c_n such that the encoding of c_n (as a string over $\{0,1\}$) has size at most $p(n)$, and c_n computes χ_A^n. Consider stage n where n is chosen so large that

$$p(n) < n^{\log n} < 2^n.$$

(The second inequality is satisfied for $n \geq 17$). Since the first inequality is satisfied, in the second for-loop there occurs a string w that encodes a circuit which computes χ_A^n. Then, by the construction, (non)membership of x_i^n to $A_{n,i}$ is determined in such a way that at least half of the circuits which are correct on $A_{n,i-1}$ are "spoiled". Since there are no more than $2^{n^{\log n}}$ potential circuits of size smaller than $n^{\log n}$, after the spoiling process has taken place 2^n times, by the second inequality, no circuit can remain to compute $A_{n,2^n}$ correctly.

Now consider the complexity of the set A. To decide whether $x \in A$ or not, $|x| = n$, it suffices to perform stage n of the construction which consists of two nested exponentially bounded for-loops where the inner loop can be performed reusing the same space. It follows that A can be computed with space $2^{O(n)}$ (and time $2^{2^{O(n)}}$). Hence, $A \in$ EXPSPACE. □

New proof techniques seem to be necessary to improve Theorem 2.5 from "EXPSPACE" to "EXPTIME" because there exist relativizations relative to which every language in EXPTIME has (relativized) polynomial size circuits (cf. Heller, 1983; Wilson, 1983). On the other hand, a proof that each set in EXPTIME has polynomial size circuits would have severe consequences: it would follow EXPTIME = PH = Σ_2^p (Karp and Lipton, 1980). Then from the time hierarchy theorem (see Hopcroft and Ullman, 1979) it follows that $P \neq \Sigma_2^p$ and thus $P \neq NP$ (since $P \neq$ EXPTIME).

2.3. GENERATING CIRCUITS

Up to now we considered (sequences of) circuits as "acceptors" of languages. Next we investigate the situation that circuits "generate" a language.

<u>Definition 2.6</u>. (Yap, 1983; Schöning, 1984) A set $A \subseteq \{0,1\}^*$ has <u>polynomial size generators</u> if for a polynomial p and each $n \geq 0$ there is circuit c of size at most $p(n)$ having $k \leq p(n)$ inputs such that there are $s, r_1, \ldots, r_n$ with

$$A_{=n} = \{res_c^{r_1}(x_1,\ldots,x_k)\ldots res_c^{r_n}(x_1,\ldots,x_k) \mid x_1,\ldots,x_k \in \{0,1\} \text{ and } res_c^{s}(x_1,\ldots,x_k)=1\}.$$

Observe that in this definition s only serves as a "domain indicator", such that it is possible to apply the definition also to sets A where for some n, $A_{=n}$ is empty.

It is easy to see that each language A having polynomial size circuits has polynomial size generators: Let $n \geq 0$ and let c be a circuit with n inputs such that $f_c = \chi_A^n$. Then c satisfies the definition of a generating circuit for $A_{=n}$ choosing $k=n$, $s=n+q$, $r_1=1,\ldots,r_n=n$.

It is an open question whether the inverse assertion is true. We will show in the following that there exists a strong link between this question and $P =? NP$.

<u>Corollary 2.7</u>. There exist sets in EXPSPACE not having polynomial size generators.

<u>Proof</u>. This follows with the same technique as in Theorem 2.5. Each evaluation of $f_c(x)$ has to be substituted by a test whether x can be "generated" by circuit c with an appropriate input. For this test an additional for-loop is needed which cycles through all potential inputs to c. This modification does not change the fact that $A \in$ EXPSPACE. □

2.4. SIMULATION OF TURING MACHINES BY CIRCUITS

In chapter 2.1 it was shown that sets with small circuit-size complexity might have arbitrarily high Turing machine complexity. The circuit-size complexity never exceeds, say, 2^n. But, on the other hand, small (deterministic) Turing machine complexity does imply small circuit-size complexity.

<u>Theorem 2.8</u>. For each set $L \in DTIME(t(n))$, $CS_L = O(t^2(n))$.

<u>Proof</u> (sketch). Let M be a t(n)-time bounded deterministic Turing machine that accepts L. Let Q be M's set of states, and let A be the work-tape symbols, $\{0,1,b\} \subseteq A$, where b is the blanc symbol, $A \cap Q$ is empty. Suppose M has k tapes. Let $n \geq 0$ be given. Each configuration of M on an input $x = x_1 \ldots x_n$ can be represented as a string of size t(n) over the alphabet $D = (A \cup (Q \times A))^k$ in an obvious way. E.g., the initial configuration (of a 1-tape Turing machine) is represented by $(q_0,x_1)x_2 \ldots x_n b^{t(n)-n}$ where $q_0 \in Q$ is the initial state. From M's transition function a finite function $f : D^3 \to D$ can be obtained such that the i-th symbol of the follow configuration of a given configuration $c_1c_2 \ldots c_{t(n)}$ can be computed as $f(c_{i-1},c_i,c_{i+1})$. This function f can be realized by a finite circuit where the elements of D are represented as 0-1-strings (say, of length $\lceil 1 + \log |D| \rceil$). Then it is possible to simulate computations of M by circuits of the following form:

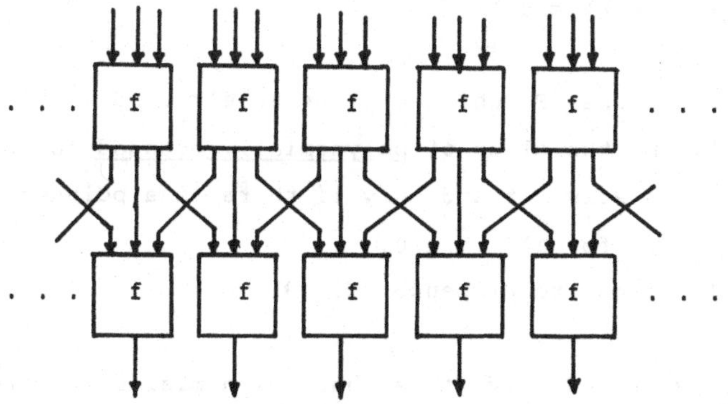

(k-th configuration)

((k+1)-th config.)

Some additional circuitry is needed to connect the input $x = x_1 \ldots x_n \in \{0,1\}^n$ to the layer of the circuit representing the initial configuration. Further, from the t(n)-th configuration a signal has to be drawn out that indicates whether M accepts. Such a circuit obviously computes χ_L^n, and the form of the circuit immediately shows that its size is $O(t^2(n))$, thus $CS_L(n) = O(t^2(n))$. □

The proof of Theorem 2.8 has been adapted from Ladner's result that the problem of deciding whether a given circuit computes 1 on a given input is P-complete under log-space reducibility (cf. Ladner, 1975a). Fischer and Pippenger (see Fischer, 1974) have shown that the bound in Theorem 2.8 can be improved to $O(t(n) \cdot \log t(n))$.

Corollary 2.9. Each set in P has polynomial size circuits.

Corollary 2.10. No $\leq_m^P$- (or even $\leq_T^P$-)hard set for EXPSPACE can have polynomial size circuits.

Proof. This follows from Theorem 2.5, together with the observation that a construction like in the proof of Theorem 2.8 can be used to simulate a polynomial-time reduction by a polynomial size circuit. □

Now we introduce an important notation which leads to a characterization of the sets having polynomial size circuits.

Definition 2.11. (Karp and Lipton, 1980) For a class of sets C and a class of functions F from N to Σ^* define C/F as the class of sets $A \subseteq \Sigma^*$ for which there exists a set $B \in C$ and a function $h \in F$ such that $A = \{ x \mid \langle x, h(|x|) \rangle \in B \}$.

We fix a class of functions F which will be solely used in the following. Let F = Poly be the class of polynomially bounded functions from N to Σ^*, i.e. $h \in$ Poly if and only if there is a polynomial p such that $|h(n)| \leq p(n)$ for all $n \geq 0$.

The following result is due to Pippenger (1979).

Theorem 2.12. $A \in$ P/Poly if and only if A has polynomial size circuits.

Proof. Let $A \in P/Poly$. Then there is a $B \in P$ and a function $h \in Poly$ such that $A = \{ x \mid \langle x, h(|x|) \rangle \in B \}$. Suppose $B = L(M)$ for a deterministic, polynomial-time Turing machine M. Similar to the proof of Theorem 2.8, there is a circuit of polynomial size with n inputs (depending on M and h) that outputs "1" if and only if $\langle x, h(|x|) \rangle \in B$, i.e. it computes χ_A^n.

Conversely, suppose A is a set having polynomial size circuits. For each $n \geq 0$ let $h(n) \in \Sigma^*$ be the encoding of the circuit that computes χ_A^n. Clearly h is polynomially bounded, i.e. $h \in Poly$. Then $A = \{ x \mid \langle x, h(|x|) \rangle \in B \}$ for a set $B \in P$ where

$\langle x, y \rangle \in B$ iff y is encoding of a circuit c with $|x|$ inputs and $f_c(x) = 1$.

∎

Observe that from this proof, B may be viewed as the "circuit interpreter" and h(n) as (the encoding of) the n-th circuit that computes χ_A^n.

The following theorem connects the P-NP-question with the question whether small generating circuits can always be transformed into small "accepting" circuits. The interesting point about this result is that we have a version of the P-NP-question within the context of circuit complexity, and circuits are purely combinatorial objects independent of any particular machine model.

Theorem 2.13. $A \in NP/Poly$ if and only if A has polynomial size generators.

Proof. Suppose $A \in NP/Poly$. Then there is a $B \in NP$ and an $h \in Poly$ such that $A = \{ x \mid \langle x, h(|x|) \rangle \in B \}$. By the quantifier characterization of the polynomial hierarchy (Stockmeyer, 1977) it follows that there is a deterministic and polynomial-time bounded Turing machine M and a polynomial p such that

$$A = \{ x \mid (\exists y, |y| = p(|x|)) \ \langle x, h(|x|), y \rangle \in L(M) \}.$$

Similar to Theorem 2.12, for each $n \geq 0$ there is a circuit c that depends on M and h having $n + p(n)$ inputs. The size of c is polynomial in n, and for each x, $|x| = n$, and y, $|y| = p(n)$,

$$f_c(xy) = 1 \text{ if and only if } \langle x, h(|x|), y\rangle \in L(M).$$

Now, c is a generating circuit for $A_{=n}$ according to Definition 2.11 taking $s=n+q$, $r_1=1, \ldots, r_n=n$. This shows that A has polynomial size generators.

Conversely, suppose A has polynomial size generators. Then for a polynomial p and each $n \geq 0$ there is a circuit c of size at most $p(n)$ with $k \leq p(n)$ inputs such that for some $s, r_1, \ldots, r_n$,

$$A_{=n} = \{ res_c^{r_1}(x) \ldots res_c^{r_n}(x) \mid x \in \{0,1\}^k \text{ and } res_c^s(x) = 1 \}.$$

Choose a function $h : N \to \Sigma^*$ such that $h(n)$ is a suitable encoding of $c, k, s, r_1, \ldots, r_n$. Then $h \in$ Poly. Furthermore, $A = \{ x \mid \langle x, h(|x|)\rangle \in B \}$ for an appropriate language B,

$\langle x,y\rangle \in B$ iff y is encoding of $c, k, s, r_1, \ldots, r_n$ such that there exists an input $a_1 \ldots a_k \in \{0,1\}^k$ for the circuit c with $res_c^s(a_1 \ldots a_k) = 1$ and $x = res_c^{r_1}(a_1 \ldots a_k) \ldots res_c^{r_n}(a_1 \ldots a_k)$.

Now it is easy to see that $B \in NP$ (the nondeterminism is used here to "guess" the input $a_1 \ldots a_k$ for the circuit). The above chararcterization of A shows that $A \in NP/Poly$. □

The backward direction in Theorem 2.13 was first proved by Yap (1983) whereas the foreward direction was stated as an open question, and was solved in (Schöning, 1984).

The question whether P/Poly = NP/Poly (i.e. whether polynomial size generators can always be substituted by polynomial size "accepting" circuits) is open. A (positive or negative) solution to this question would have important consequences for other unsolved problems in complexity theory. If P/Poly ≠ NP/Poly then, immediately from the definition, P ≠ NP. On the other hand, if P/Poly = NP/Poly then this implies NP ⊆ P/Poly (because NP ⊆ NP/Poly). This means with Theorem

2.12 that each set in NP has polynomial size circuits - a very unlikely event. It implies that the polynomial-time hierarchy "collapses" to the second level, i.e. $PH = \Sigma_2^P$ (Karp and Lipton, 1980). We will show this and related results in a later section.

The following recent result by Hartmanis and Yesha (1983) makes an assertion about the possible existence of sets having polynomial size circuits in PSPACE - P. It is shown that this is equivalent to EXPTIME $\neq$ EXPSPACE.

<u>Theorem 2.14</u>. (Hartmanis and Yesha, 1983) EXPTIME $\neq$ EXPSPACE if and only if PSPACE $\cap$ P/Poly $\neq$ P.

<u>Proof</u>. Suppose there is a set A in EXPSPACE - EXPTIME. A simple padding argument (see Book, 1974) shows that the set

$$\text{TALLY}(A) = \{ 0^n \mid \text{the binary encoding of } n \text{ has the form } 1w \text{ and } w \in A \}$$

is in PSPACE - P. Since TALLY(A) $\subseteq \{0\}^*$, we have TALLY(A) $\in$ P/Poly.

Conversely, suppose there is a set A in (PSPACE $\cap$ P/Poly) - P. Since A $\in$ P/Poly there is a set B $\in$ P and a polynomial p such that for each $n \geq 0$ there is a string w of size p(n) with $A_{=n} = \{ x \mid \langle x,w\rangle \in B\}_{=n}$. For each $n \geq 0$ let w_n be the first such string in lexicographical order having the above property. Now we encode the set $\{ w_n \mid n \geq 0 \}$ "bitwise" into a language C as follows.

$$C = \{ \langle 0^n, 0^k\rangle \mid \text{the k-th symbol of } w_n \text{ is "1"} \}.$$

This set C is in PSPACE because, on input $\langle 0^n, 0^k\rangle$, it is possible to enumerate (on the same space) all strings w of length p(n) until the the first w is found that satisfies for all strings x, $|x| = n$,

$$x \in A \text{ if and only if } \langle x,w\rangle \in B.$$

After this, it has to be checked whether the k-th symbol of this particular w (i.e. w_n) is 1. By the fact that A $\in$ PSPACE and B $\in$ P the above procedure can be performed in polynomial space.

On the other hand, C is not in P by the fact that $A \notin P$ and $A \in P(C)$. (To see that $A \in P(C)$, consider the following oracle machine that, on input x queries its oracle C about the strings $\langle 0^{|x|}, 0^k \rangle$, $k = 1, \ldots, p(|x|)$. By this, the machine can determine $w_{|x|}$ and then check whether $\langle x, w_{|x|} \rangle \in B$.)

Now we know that $C \in$ PSPACE − P where C is a tally set. A similar argument as in the first part of this proof (cf. Book, 1974) shows that the "compressed" version of C,

$$D = \{ \langle x, y \rangle \mid x \text{ is the binary encoding of } n,\ y \text{ is the binary encoding of } k, \text{ and } \langle 0^n, 0^k \rangle \in C \},$$

is in EXPSPACE − EXPTIME. □

2.5. P-SELECTIVE SETS

The interesting class of p-selective languages is introduced. We will show that all languages of this class have polynomial-size circuits. The p-selective sets have been studied intensively by Selman (1979; 1982a; 1982b). Selman has translated the recursion-theoretic notion of "semi-recursive" sets by Jokusch (1968) into the context of polynomial-time bounded computations.

Definition 2.15. (Selman, 1979) A language A is called p-selective if there is a 2-placed function f (the selector for A) computable in polynomial time such that for each $x, y \in \Sigma^*$,

(i) $f(x, y) = x$ or $f(x, y) = y$,

(ii) if $x \in A$ or $y \in A$ then $f(x, y) \in A$.

Selman used p-selective sets in several ways to distinguish various types of polynomial-time bounded reducibilities - especially on NP. It is known that $\leq^P_m$-complete sets for NP cannot be p-selective unless $P = NP$. (Note that each set in P is trivially p-selective). Selman's conjecture is that the $\leq_T^P$-complete sets for NP might contain p-selective sets, and hence, p-selective sets distinguish $\leq_m^P$- and $\leq_T^P$-completeness for NP (cf. Selman 1979; 1982a; 1982b). Theorem 2.17 will tell us that Selman's conjecture is false if the polynomial-time hierarchy does not collapse to its second level Σ_2^P. This follows from the result that p-selective sets have polynomial size circuits. This was first proved by Ko (1982) in a somewhat general form.

First we need a technical lemma.

<u>Lemma 2.16</u>. Let A be a p-selective set, and let f be the selector function from Definition 2.15. Then for each $n \geq 0$ there is a set $C \subseteq A_{=n}$, $|C| \leq n+1$, such that

$$A_{=n} = \{ x \mid |x|=n \text{ and for some } y \in C \text{ holds } f(x, y) = x \text{ or } f(y, x) = x \}.$$

<u>Proof</u>. Let $n \geq 0$ be given. Define a relation R_f^n on Σ^n such that $x R_f^n y$ if and only if $f(x, y) = x$ or $f(y, x) = x$. Now for each pair of strings (x, y) of size n either $x R_f^n y$ or $y R_f^n x$ (or both). Thus,

$$|R_f^n \cap (A_{=n})^2| \geq |A_{=n}|^2/2.$$

Suppose $A_{=n}$ is not empty (otherwise we are ready). Then there exists a $y_0 \in A_{=n}$ with

$$|\{ x \mid x R_f^n y_0 \}| \geq |A_{=n}|/2.$$

Let X_0 be the set $\{ x \mid x R_f^n y_0 \}$. By property (ii) of the selector function f, $X_0 \subseteq A_{=n}$. By the fact that

$$|R^n{}_f \cap (A_{=n} - X_0)^2| \geq |A_{=n} - X_0|^2/2,$$

there exists a $y_1 \in A_{=n} - X_0$ (if not yet empty) such that

$$|\{ x \in X_0 \mid x R_f^n y_1 \}| \geq |A_{=n} - X_0|/2.$$

Continuing by this argument the remaining set becomes empty after at most n steps because $|A_{=n}| \leq 2^n$. Let $y_0, y_1, \ldots, y_k$, $k \leq n$, be the strings obtained in this way. Then the set $C = \{y_0, y_1, \ldots, y_k\}$ has the desired property. □

Using Lemma 2.16 which states that for each n always a "small" set C can be found that encodes all "information" about $A_{=n}$, it is not hard to prove the following theorem.

Theorem 2.17. (Ko, 1983) Each p-selective set is in the class P/Poly, i.e. each p-selective set has polynomial size circuits.

Proof. Let A be p-selective, and let f be a suitable selector function. For each $n \geq 0$ let $h(n) \in \Sigma^*$ be the encoding of the set $C = \{y_0, \ldots, y_k\}$, $k \leq n$, provided by Lemma 2.16. Clearly, $h \in$ Poly. Now let B be the following language:

$$B = \{ \langle x,w \rangle \mid w \text{ is encoding of a set of strings } C \text{ of size } |x| \text{ and for some } y \in C,\ f(x, y) = x \text{ or } f(y, x) = x \}.$$

Clearly, $B \in P$ and $A = \{ x \mid \langle x,h(|x|)\rangle \in B \}$. Thus it follows that $A \in$ P/Poly. □

Ko (1983) proved this result for a little larger class, the "weakly p-selective" sets (which include polynomial left cuts). By results to be proved later, neither $\leq^P_m$- nor $\leq^P_T$-complete sets can have polynomial size circuits unless $PH = \Sigma^P_2$. Hence it is unlikely that Selman's conjecture turns out to be true.

In chapter 5 we will turn back to the p-selective sets and give a classification in terms of "lowness". There the results developed here will be needed.

CHAPTER 3

PROBABILISTIC ALGORITHMS

In probabilistic algorithms some computational steps may depend on the outcome of an (ideal) random generator. Both, the run time and the output of such a probabilistic algorithm now become random variables. Their probability distributions may depend on the input of the algorithm. In this chapter we will define and study some complexity classes which are defined in terms of probabilistic algorithms (Turing machines) - we require that they are, in a sense, polynomially time bounded. Further we study their relationships to classes like P, NP, PH and P/Poly.

It should be observed that the model of a probabilistic algorithm studied here has nothing to do with the investigation of the probabilistic behavior of <u>deterministic</u> algorithms when assuming some reasonable probability distribution on the set of possible inputs.

Probabilistic algorithms can be classified according to various criteria. In the next section we introduce several types of probabilistic algorithms, and hence, classes of languages defined in terms of these algorithms. One possible distinction is in "Monte Carlo" and "Las Vegas" type algorithms (cf. Zachos, 1982). Another important property of probabilistic algorithms is that the probability of not producing a correct output can be made arbitrarily small by "iterating" the algorithm several times on the same input with independent random choices. It will be shown that exactly this property gives the link to circuit-size complexity.

3.1. MONTE CARLO OR LAS VEGAS? CLASSIFICATIONS OF PROBABILISTIC ALGORITHMS

A <u>probabilistic Turing machine</u> is just a nondeterministic Turing machine where each nondeterministic choice is considered as a random experiment, each possible branch having equal probability. For simplicity (and without loss of generality) we assume that each nondeterministic branch has two possible outcomes - each assigned probability 1/2. Furthermore, we distinguish 3 types of final states (and hence, final configurations): 1-states (or, accepting states), 0-states (or, rejecting states), and ?-states (or, "don't know"-states).

Since each nondeterministic choice has probability 1/2, each nondeterministic computation of length t has probability 2^{-t}. The outcome (i.e. the type of final state) of a probabilistic Turing machine M on input x now becomes a random variable, denoted $M(x)$, whose range is the set $\{0,1,?\}$. Let $\text{Prob}[\ M(x) = a\ ]$ be the probability that M on input x halts in an a-state, $a \in \{0,1,?\}$.

Several classes of languages can be defined in terms of polynomial time bounded probabilistic machines (cf. Gill, 1977; Adleman and Manders, 1977).

<u>Definition 3.1</u>. The class PP (probabilistic polynomial time) consists of all languages $A \subseteq \Sigma^*$ for which there exists a probabilistic and polynomial time bounded Turing machine M such that for each $x \in \Sigma^*$, $\text{Prob}[\ M(x) = \chi_A(x)\ ] > 1/2$.

It can be shown that $NP \cup co\text{-}NP \subseteq PP \subseteq PSPACE$ (Gill, 1977). None of these inclusions is known to be proper. Obviously, if $P \neq NP$ (which is general belief) then $P \neq PP$. It is not known whether PP is included in the polynomial hierarchy PH. PP is closed under complement, and the following language

$$\#SAT = \{ \langle i, F\rangle \mid F \text{ is a Boolean formula with at least } i \text{ different satisfying assignments} \}$$

has been shown to be PP-complete by extending Cook's proof that SAT is NP-complete (Simon, 1975; Gill, 1977). We will come back to this language in a later chapter.

The Turing machines allowed in Definition 3.1 are of the "Monte Carlo" type. That is, they are allowed to "lie" with small probability. (On an input $x \in A$ there is a chance for the output "0", and conversely on input $x \notin A$ there is a chance for the output "1"). This is not allowed for the "Las Vegas" type algorithms (only "?" is allowed but not a wrong answer).

A more restricted class than PP, but still of the Monte Carlo type, is BPP.

Definition 3.2. The class BPP (bounded error probabilistic polynomial time) consists of all languages $A \subseteq \Sigma^*$ for which there exists a probabilistic polynomial time bounded Turing machine M and a number e, $0 < e < 1/2$, such that for each $x \in \Sigma^*$, $Prob[\ M(x) = \chi_A(x)\ ] > 1/2 + e$.

It is obvious that $P \subseteq BPP \subseteq PP$. On the other hand, PP contains NP but this is not known to hold (and unlikely) for the class BPP. Further, by a recent result by Gács (stated in Sipser, 1983) and independently obtained by Lautemann (1983), BPP is contained in the polynomial hierarchy, more specifically, $BPP \subseteq \Sigma_2^P \cap \Pi_2^P$. We will prove an even stronger result in chapter 5.

There exists an interesting "one-sided" version of BPP.

Definition 3.3. The class R (random polynomial time) consists of all languages $A \subseteq \Sigma^*$ for which there exists a probabilistic and polynomial time bounded Turing machine M such that for each $x \in \Sigma^*$,

(i) $x \in A \Rightarrow Prob[\ M(x) = 1\ ] > 1/2$,

(ii) $x \notin A \Rightarrow Prob[\ M(x) = 1\ ] = 0$.

In Definition 3.3 (i) the number 1/2 can be altered to any constant d, $0 < d < 1$, without changing the defined class. This is true because iterating the "random experiment" M(x) k times independently, and finally outputting "1" if $M(x) = 1$ occures in at least one trial, yields

$$x \in A \Rightarrow Prob[\ M_k(x) = 1\ ] > 1 - (1-d)^k,$$

where M_k is the above descibed "k-iterated" probabilistic machine. By an appropriate choice of k, it is possible to garanty $1 - (1-d)^k > 1/2$, even stronger $1 - (1-d)^k > 1/2 + e$ for any constant e, $0 < e < 1/2$.

This argument shows that $R \subseteq BPP$, and by the fact that BPP is closed under complement, co-R $\subseteq$ BPP. Further, immediately from the definition of R, $P \subseteq R \subseteq NP$. The importance of the class R stems from this relation to the P-NP-question. Further, Rabin (1976) has shown that the problem of deciding whether (the binary encoding of) a given number is composite is in the class R. A similar probabilistic test has been presented by Solovay and Strassen (1977).

Let ZPP (zero error probabilistic polynomial time) be the class $R \cap$ co-R. From $P \subseteq R \subseteq NP$ follows that $P \subseteq ZPP \subseteq NP \cap$ co-NP. The name "zero error" stems from the following easy to prove characterization of ZPP (Gill, 1977; Zachos, 1982): $A \in ZPP$ if and only if there is a probabilistic and polynomial time bounded Turing machine M such that for each $x \in \Sigma^*$,

(i) $x \in A \Rightarrow Prob[\ M(x) = 0\ ] = 0$ and $Prob[\ M(x) = 1\ ] > 1/2$,
(ii) $x \notin A \Rightarrow Prob[\ M(x) = 1\ ] = 0$ and $Prob[\ M(x) = 0\ ] > 1/2$.

Thus, ZPP-type algorithms belong to the "Las Vegas"-type, wheras R is, in a sense, a mixture: On input $x \in A$ an R-machine may "lie" with small probability, but not on input $x \notin A$.

Further, it can be shown that ZPP is exactly the class of sets that can be recognized by probabilistic Turing machines having <u>average</u> polynomial running time (Gill, 1977).

It is not known whether the classes ZPP, R, BPP possess complete sets. There exists certain evidence that this is not the case (Adleman, 1978; Sipser, 1982).

The diagram on the next page summarizes all known inclusions between the defined probabilistic classes, the classes of the polynomial-time hierarchy, and PSPACE.

The most interesting open questions in this context are:

$PP \subseteq PH$? $NP \subseteq BPP$ or $BPP \subseteq NP$ or neither of them ?

The assumption $NP \subseteq BPP$ has some very unlikely consequences (which will all be shown later) like $NP \subseteq P/Poly$ (since $BPP \subseteq P/Poly$, see Theorem 3.7), $NP = R$ (Theorem 3.6), and $PH = \Sigma_2^P$ (see remarks after Theorem 4.29).

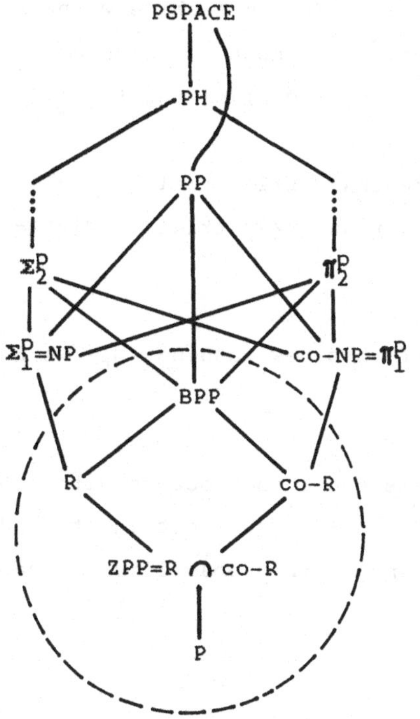

3.2. ITERATION OF PROBABILISTIC ALGORITHMS

The broken line in the above diagram surrounds those classes which may be considered as computational feasible. The reason is that for these classes the probability of producing an incorrect output can be made "arbitrarily" small (e.g. smaller than the probability for the occurence of a hardware error) by reapeating the algorithm on the same input a certain number of times with independent random choices. We already described how to do this in the case of the class R. Iterating the algorithm t times increases the probability for a correct answer from d to $1 - (1-d)^t$. Hence for $d = 1/2$ as in Definition 3.3 we achieve the probability $1 - 2^{-t}$. For the class ZPP the situation is very similar.

Somewhat more complicated is the situation with the class BPP because we cannot always trust the outcome of a BPP-type algorithm, we may get errors in both ways. Here the following procedure helps:

Let M be a probabilistic Turing machine witnessing that $A \in BPP$. Consider a machine M_t which on input x simulates M on x t times (t is odd and may depend on x) and gives the following output:

$$M_t(x) = \begin{cases} 1, & \text{if for more than } t/2 \text{ trials, } M(x) = 1, \\ 0, & \text{if for more than } t/2 \text{ trials, } M(x) = 0, \\ ?, & \text{otherwise.} \end{cases}$$

To analyse the behavior of this new machine M_t we need the following lemma.

<u>Lemma 3.4</u>. Let E be some event that occurs with probability at least $1/2 + e$, $0 < e < 1/2$. Then E occurs within t independent trials (t odd) more than $t/2$ times with probability at least

$$1 - \tfrac{1}{2}\cdot(1 - 4e^2)^{t/2} .$$

(That is, for fixed e and $t \to \infty$, this probability goes exponentially to 1).

<u>Proof</u>. Let q_i be the probability that E occurs exactly i times in t independent trial. By the binomial distribution (cf. Feller, 1957),

$$\begin{aligned} q_i &= \binom{t}{i}\cdot(1/2 + e)^{i}\cdot(1/2 - e)^{t-i} \\ &\le \binom{t}{i}\cdot(1/2 + e)^{i}\cdot(1/2 - e)^{t-i}\cdot\left(\frac{1/2 + e}{1/2 - e}\right)^{t/2 - i} \qquad \text{for } i \le t/2 \\ &= \binom{t}{i}\cdot(1/4 - e^2)^{t/2} . \end{aligned}$$

Thus we get for the probability that E occurs more than $t/2$ times,

$$\begin{aligned} p &= 1 - \sum_{i=0}^{[t/2]} q_i \\ &\ge 1 - \sum_{i=0}^{[t/2]} \binom{t}{i}\cdot(1/4 - e^2)^{t/2} \\ &= 1 - 2^{t-1}\cdot(1/4 - e^2)^{t/2} \\ &= 1 - \tfrac{1}{2}\cdot(1 - 4e^2)^{t/2} . \quad \square \end{aligned}$$

The bound given in Lemma 3.4 is sharper than the "Chernoff bound" (Chernoff, 1952; cf. Erdős and Spencer, 1974, p.18). Using this lemma, we will see that the error probability in BPP-type algorithms can be made "arbitrarily" small.

<u>Theorem 3.5</u>. Let $A \in BPP$. Then, for each polynomial q there is a probabilistic and polynomial time bounded Turing machine M such that for each $x \in \Sigma^*$,

$$\text{Prob}[\ M(x) = \chi_A(x)\ ] \ > \ 1 - 2^{-q(|x|)}.$$

<u>Proof</u>. Let $A \in BPP$. Then there is a probabilistic polynomial time bounded Turing machine M and a constant e, $0 < e < 1/2$, such that for each $x \in \Sigma^*$,

$$\text{Prob}[\ M(x) = \chi_A(x)\ ] \ > \ 1/2 + e\ .$$

Consider the "t-iterated" machine M_t described above. Now, t has to be chosen such that

$$1 - \frac{1}{2}\cdot(1 - 4e^2)^{t/2} \ \geq \ 1 - 2^{-q(|x|)} ,$$

or equivalently,

$$2\cdot(1/(1 - 4e^2))^{t/2} \ \geq \ 2^{q(|x|)} .$$

It follows that it suffices to take $t = t(|x|) = c\ q(|x|)$ where

$$c = \frac{2}{\log\ (1/(1 - 4e^2))} .$$

The table on the next page illustrates the relationship between c and e. Now, this machine M_t is $O(tp(n)) = O(q(n)p(n))$ time bounded where p is the time bound of M, and it has the desired properties. □

e	c
0.05	137.94
0.1	33.96
0.15	14.7
0.2	7.95
0.25	4.82
0.3	3.11
0.35	2.06
0.4	1.36
0.45	0.83

Using the fact that BPP-machines can have exponentially small error probabilities, it is not hard to prove the following theorem. Other applications of Theorem 3.5 will be given in the next section.

Theorem 3.6. (Ko, 1982) If $NP \subseteq BPP$ then $NP = R$.

Proof. Suppose $NP \subseteq BPP$, hence $SAT \in BPP$. By the NP-completeness of SAT it suffices to show that this implies $SAT \in R$. SAT has a certain "self-reducibility" property (discussed in more detail in section 4.5) which means that a formula $F(x_1,\ldots,x_n)$ is satisfiable if and only if $F_0 := F(0,x_2,\ldots,x_n)$ is satisfiable or $F_1 := F(1,x_2,\ldots,x_n)$ is satisfiable. (That is, the decision whether the instance F is satisfiable can be "reduced" to two "smaller" instances F_0, F_1 where "smaller" means fewer variables here). We may assume an encoding of SAT such that (the encoding of) each formula may be "padded" with redundant information such that the size of all formulas in the tree

```
            F
          /   \
       F0       F1
      /  \     /  \
   F00   F01 F10   F11
   / \   / \  / \   / \
         . . . . . . .
```

is equal to the size of F. This assumption makes the following calculation somewhat easier but is not essential. By Theorem 3.5 we may choose a probabilistic and polynomial time bounded Turing machine M for SAT such that for each $x \in \Sigma^*$,

$$\text{Prob}[\ M(x) = \chi_{SAT}(x)\ ] \ > \ 1 - 2^{-|x|} .$$

The following algorithm will be shown to witness that $SAT \in R$.

```
begin
  input  x ; {x is encoding of some formula  F(x_1,...,x_n), n ≤ |x|}
  if  M(x) = 0  then  reject ;
  G := F ;
  for  i := 1 to  n  do
    begin
      if M(<G_0>)=1 then
        begin
          G := G_0 ;
          a_i  := 0 ;
        end
      else if M(<G_1>)=1 then
          begin
            G := G_1 ;
            a_i  := 1 ;
          end
        else  reject ;
    end;
  if  F(a_1,...,a_n) = 1  then  accept                    (*)
                          else  reject ;
end.
```

Observe that this algorithm M' will never accept unsatisfiable formulas because of the final test in line (*). Hence

$$x \notin SAT \ \Rightarrow\ \text{Prob}[\ M'(x) = 1\ ] = 0 .$$

On the other hand, if $x = \langle F \rangle$ is satisfiable, in the for-loop a satisfying assignment $a_1,\ldots,a_n$ will be constructed with probability

$$(1 - 2^{-|x|})^n \geq (1 - 2^{-|x|})^{|x|} > 1/2 \quad \text{for} \quad |x| > 1.$$

Hence, SAT $\in$ R and therefore NP = R. □

3.3. PROBABILISTIC CLASSES AND POLYNOMIAL SIZE CIRCUITS

Using Theorem 3.5, a link between probabilistic classes and the notion of polynomial size circuits can be established.

<u>Theorem 3.7</u>. Each set in BPP has polynomial size circuits.

<u>Proof</u>. Let $A \in$ BPP, then we choose, using Theorem 3.5, a probabilistic polynomial time bounded Turing machine M such that for each x, $|x| = n$

$$\text{Prob}[\ M(x) = \chi_A(x)\ ] > 1 - 2^{-2n} .$$

Let p be a polynomial bounding the run time of M. We may assume that each computation of M on an input of size n has <u>exactly</u> length $p(n)$. Each such nondeterministic computation can be represented by a string y $\Sigma^{p(n)}$. For each x, $|x| = n$, there are at most $2^{p(n)-2n}$ y's not leading to the correct output. Since there are 2^n strings x of size n, there exist at least

$$2^{p(n)} - 2^n \cdot 2^{p(n)-2n} = 2^{p(n)} \cdot (1 - 2^{-n}) > 1$$

strings y that describe correct computations for <u>all</u> inputs x, $|x| = n$. For each n, let $h(n)$ be the first such y, $|y| = p(n)$, according to lexicographical order. Then $A = \{\ x \mid \langle x, h(|x|)\rangle \in B\ \}$ where B is defined as follows.

$\langle x,y\rangle \in B$ iff $|y| = p(|x|)$ and M on input x, using the nondeterministic computation y, accepts.

It is clear that $B \in P$ and $h \in$ Poly, hence $A \in$ P/Poly. □

Corollary 3.8. All sets in the classes ZPP, R, co-R have polynomial size circuits.

The fact that the sets in R have polynomial size circuits was first shown by Adleman (1978). The above proof is inspired by the work of Bennett and Gill (1981). Observe that the broken line in the diagram just surrounds those classes shown to be in P/Poly.

Corollary 3.9. P $\neq$ BPP implies EXPTIME $\neq$ EXPSPACE.

Proof. This follows from Theorem 3.7 and Theorem 2.14. ◻

Hartmanis and Yesha (1983) pointed out that Corollary 3.8 is the only known "upward separation" result. All other such implications known in the literature are of the "downward separation" type, which means that from the inequality of two "high" classes the inequality of two "low" classes follows. Some examples:

EXPTIME $\neq$ NEXPTIME implies P $\neq$ NP (Book, 1974),

EXPTIME $\neq$ EXPSPACE implies P $\neq$ PSPACE (Book, 1974),

$\Sigma_k^p \neq \Sigma_{k+1}^p$ for some $k \geq 1$ implies P $\neq$ NP (Stockmeyer, 1977).

Using the proof method of Theorem 3.7, a characterization of BPP can be obtained which shows that BPP can be expressed as a strong version of P/Poly. This is in spirit similar to the work of Zachos and Heller (1984).

Theorem 3.10. The following are equivalent:

(i) $A \in$ BPP,

(ii) For each polynomial q there is a set $B \in P$ and a polynomial p such that for each $n \geq 0$, if y is chosen uniformly at random, $|y| = p(n)$, then with probability at least $1 - 2^{-q(n)}$,

$$A_{=n} = \{ x \mid \langle x,y\rangle \in B \}_{=n} .$$

<u>Proof</u>. That (ii) implies (i) is trivial. Conversely, suppose $A \in BPP$. Then consider the proof idea of Theorem 3.7. Instead of choosing the probability for a correct output of the machine M as $1 - 2^{-2n}$, we now choose $1 - 2^{-q(n)-n}$. Then the rest of the proof goes trough where the probability of a string y that is correct for all inputs of size n is then at least $1 - 2^{-q(n)}$. □

This characterization of BPP will be used in chapter 5 to demonstrate the "lowness" of BPP.

CHAPTER 4

SPARSE SETS

A set will be called sparse (formal definitions follow) if it has, in
 sense, only few elements, that is, its "information content" is low.
he question whether sparse sets in NP - P (in particular, NP-
omplete sets) exist was first studied by Berman and Hartmanis (1977).
he existence of a sparse NP-complete set would contradict the so-called
erman-Hartmanis-conjecture. This conjecture states that all NP-complete
ets are pairwise polynomially isomorphic, that is, all NP-complete sets
re "encodings" of the same set, say SAT. Berman and Hartmanis show
hat all "known" NP-complete sets are polynomially isomorhic to SAT.
pecifically, they show that all known NP-complete sets are "invertibly
addable" (defined later), and they further show that an NP-complete set
s invertibly paddable if and only if it is polynomially isomorphic to
AT (this is a modification of the Cantor-Bernstein construction).

The Berman-Hartmanis conjecture trivially implies $P \neq NP$ because P
ontains all finite sets, and no infinite set (like the known NP-complete
ets) can be (polynomially) isomorphic to a finite set. Very similar, no
parse set can be polynomially isomorphic to a non-sparse set (like the
nown NP-complete sets). Hence, naturally, Berman and Hartmanis (1977)
osed the question whether sparse NP-complete sets can exist. This
uestion has been finally settled by Mahaney (1982) who showed that
parse $\leq_m^P$-complete sets for NP cannot exist unless $P = NP$. We
ill state this and related results in section 4.1.

There are other reasons for studying sparse sets not stemming from
he Berman-Hartmanis conjecture (which now, by some recent results, is
elieved to be false, cf. Young, 1983; Hartmanis, 1983b). We will
ntroduce various approaches to "approximate" NP-complete - and other
ntractable - decision problems. One can consider algorithms that are
ast "almost everywhere" (with the exception of at most a sparse set),
r algorithms that are fast but only "almost everywhere" correct (i.e.
on-correct on at most a sparse set). We will study such algorithms for

NP-complete problems in sections 4.2 and 4.3 and obtain some negative results, in the sense that NP-complete problems cannot be solved by means of such approximation algorithms unless $P = NP$. Also, an interesting connection to the notion of "complexity cores" is drawn.

The link to the general theme of this monograph is given by the observation that sparse sets, and sets having "approximation algorithms" in the above sense all have polynomial size circuits. In section 4.4 we give a characterization of the notion of polynomial size circuits in terms of sparse (and also, tally) oracles.

Many proof techniques in the context of NP-completeness and sparseness use the "self-reducibility" property of typical NP-complete sets like SAT (discussed already in the proof of Theorem 3.6). We study self-reducibility and sparse sets again in section 4.5 and prove a general theorem that will have several applications in the following chapters.

4.1. SPARSE SETS IN NP

The study of density notions in the context of NP-completeness was initiated by Berman and Hartmanis (1977).

Definition 4.1. A set $A \subseteq \Sigma^*$ is (polynomially) sparse if for some polynomial p and each $n \geq 0$, $|A_{=n}| \leq p(n)$.

In a sense, sparse sets cannot contain much information. Observe that in general, a set over a two-letter alphabet can have 2^n strings of size n. An interesting subset of the class of sparse sets is the class of tally sets, i.e. sets over one-letter alphabets. By simple diagonalization, tally sets of arbitrarily high complexity can be constructed. Hence there exist sparse sets of arbitrary complexity. Furthermore, each sparse set S has polynomial circuit-size complexity because the circuit realization of the disjunctive normal form of χ_S^n is polynomially bounded in n.

Berman and Hartmanis (1977) show that all "known" NP-complete sets are pairwise polynomially isomorphic. (Formally, two languages A and B are polynomially isomorphic if there exists a polynomial time computable bijection f that $\leq_m^P$-reduces A to B such that f^{-1} is computable in polynomial time, too). The known Berman-Hartmanis conjecture states that <u>all</u> $\leq_m^P$-complete sets for NP are polynomially isomorphic. As discussed above, the conjecture implies $P \neq NP$. (Because finite sets are NP-complete if and only if $P = NP$).

A certain "test" for the Berman-Hartmanis conjecture is the question whether there exist sparse NP-complete sets. If they do exist then the conjecture fails because a bijective $\leq_m^P$-reduction mapping a non-sparse set to a sparse set has to map infinitely many strings to strings of non-polynomial size. And such a function cannot be computable in polynomial time.

Therefore, another (weaker) conjecture by Berman and Hartmanis was that sparse NP-complete sets for NP cannot exist. This conjecture was finally proved to be true by Mahaney (1982) after some partial results towards this conjecture obtained by P.Berman (1978) and Fortune (1979).

<u>Theorem 4.2</u>. (Mahaney, 1982) If there exist sparse $\leq_m^P$-complete (or, $\leq_m^P$-hard) sets for NP then $P = NP$.

In other investigations of this sort, weaker types of polynomial reducibilities or more general notions than "sparse" are studied. The more important results of this type are listed below.

<u>Theorem 4.3</u>. (Karp and Lipton, 1980) If there exist sparse $\leq_T^P$-hard sets for NP then $PH = \Sigma_2^P$.

Later we will show that the assumption in Theorem 4.3 is equivalent to each set in NP having polynomial size circuits (or in symbols, $NP \subseteq P/Poly$). In Theorem 4.2 it makes no difference whether the existence of sparse $\leq_m^P$-complete or $\leq_m^P$-hard sets for NP is assumed (in fact, both assumptions are equivalent). But it does make a difference for Theorem 4.3.

<u>Theorem 4.4</u>. (Mahaney, 1982) If there exist sparse $\leq_T^P$-complete sets for NP then $PH = \Delta_2^P$.

Observe that, in Theorem 4.4, both the assumption and the conclusion are stronger than in Theorem 4.3. Theorem 4.4 was strengthened by Long (1982a).

Theorem 4.5. (Long, 1982a) If there exist sparse $\leq_T^P$-hard sets for NP within Δ_2^P then $PH = \Delta_2^P$.

An interesting consequence of Theorem 4.5 is that there are no co-sparse (i.e. sets whose complements are sparse) $\leq_T^P$-complete sets for NP unless $PH = \Delta_2^P$.

Similar results for other kinds of polynomial reducibilities (which will not be discussed here) located "between" $\leq_m^P$ and $\leq_T^P$ ("truth-table reducibilities") can be found in (Ukkonen, 1983; Yesha, 1983; and Yap, 1983). For a more general reducibility ("strong nondeterministic Turing reducibility") we will strengthen Theorem 4.4 later.

Another interesting question is whether sparse sets can at all exist in NP - P (assuming $P \neq NP$) not necessarily being NP-complete. Assuming $P \neq NP$, by a "delayed diagonalization method", sets in NP - P not being NP-complete can be constructed (Ladner, 1975; Schöning, 1982). But the method does not seem to be easily modifiable to produce sparse sets in NP - P. A partial result in this direction has been obtained by Book (1974). He showed that there exist tally sets in NP - P if and only if EXPTIME $\neq$ NEXPTIME. Recently, this has been superseded by a result of Hartmanis, Immerman and Swelson (1983) (see also Hartmanis, 1983a).

Theorem 4.6. There exist sparse sets in NP - P if and only if EXPTIME $\neq$ NEXPTIME.

Proof. Suppose $A \in$ NEXPTIME - EXPTIME. Then the "spreaded" version of A, TALLY(A) (cf. section 2.2), is in NP - P.

Conversely, let $S \in NP - P$ be a sparse set. Then there is a polynomial p such that for each $n \geq 0$, $|S_{=n}| \leq p(n)$. We encode the information contained in S "bitwise" into a tally set T as follows.

$$T = \{ \langle 0^n, 0^k, 0^m, 0^i, 0^{a+1} \rangle \mid 1 \leq m \leq k \text{ and there exist}$$
$$y_1 < y_2 < \ldots < y_k, \text{ all of size } n, \text{ such that}$$

$y_1, \ldots, y_k \in S$ and the i-th symbol of y_m is $a \in \{0,1\}$ }.

By the fact that $S \in NP$, it is easy to see that $T \in NP$. In the following it is shown that $T \notin P$. Then it follows that the "compactified" version of T,

$$T' = \{ \langle n, k, m, i, a+1 \rangle \mid \langle 0^n, 0^k, 0^m, 0^i, 0^{a+1} \rangle \in T \},$$

is in NEXPTIME - EXPTIME (cf. Book, 1974).

The following algorithm computes S in polynomial time - relative to the oracle set T, in other words, the algorithm witnesses that $S \in P(T)$. Hence, the assumption that $S \notin P$ implies that $T \notin P$.

```
begin
  input x ;
  for k := p(|x|) downto 1 do
    if <0^|x|, 0^k, 0^k, 0^1, 0^1> ∈ T or
        <0^|x|, 0^k, 0^k, 0^1, 0^2> ∈ T then goto M ;
  reject ;   { S_=|x| is empty }
M: for m := 1 to k do
    begin
      z := λ ;
      for i := 1 to |x| do
        if <0^|x|, 0^k, 0^m, 0^i, 0^1> ∈ T then z := z0
                                           else z := z1 ;
      if x = z then accept ;
    end ; { of for-loop }
  reject ;
end.
```

The algorithm, in the first for-loop, determines the number of strings in $S_{=|x|}$. Having stored this number in k, the algorithm then finds out each string in $S_{=|x|}$ (bit by bit) and compares for equality with x. The reader may verify the details. □

Next we will define and investigate some classes of decision problems that can be solved approximately by efficient algorithms.

4.2. "ALMOST CORRECT" ALGORITHMS

For intractable problems for which no polynomial-time algorithm exists (or is known to exist) it can be reasonable to consider certain approximation algorithms. Since we are dealing with <u>decision</u> problems here (as opposed to <u>optimization</u> problems) it is not quite immediate what should be considered as approximation algorithm, and how to measure the approximation quality. In the following we study two versions of efficient approximations to decision problems. First we consider algorithms that are "almost correct", in the sense that they correctly decide the set in question with the exception of at most a sparse set of input instances. This situation is very similar to the "Monte Carlo" type algorithms studied in chapter 3.

<u>Definition 4.7</u>. A set A is <u>P-close</u> if there exists a set $B \in P$ such that $A \Delta B$ is a sparse set.

The closeness notion used here goes back to the work of Yesha (1983) who studied the question how close intractable sets (like NP-complete or EXPTIME-complete sets) can be to some set in P. The closeness is measured in terms of the census function of the symmetric difference.

It is not hard to see that P-close sets are included in P/Poly because, for each n, the polynomially many strings in $(A \Delta B)_{=n}$ can be encoded into a single string $h(n)$. This inclusion can be seen to be proper as follows. Let $T \subseteq \{0\}^*$ be a tally set not in P. Then consider the set $T' = \{ x \in \Sigma^* \mid 0^{|x|} \in T \}$. Clearly, $T' \in$ P/Poly. But it is not hard to show that T' cannot be P-close (otherwise $T \in P$).

Further observe, all sparse sets and all co-sparse sets are P-close (choose $B = \emptyset$ or $B = \Sigma^*$, respectively).

<u>Definition 4.8</u>. A set A is (polynomially) <u>paddable</u> if there exists a two-placed, polynomial-time computable and one-one function pad such that for each $x,y \in \Sigma^*$, $x \in A$ iff $pad(x, y) \in A$.

A set A is <u>invertible paddable</u> if, additionally, there is a polynomial-time computable function decode such that for each $x,y \in \Sigma^*$ decode(pad(x, y)) = y.

Intuitively, a set is paddable if for each instance x there are infinitely many, easily produceable, input instances (the padded versions of x) behaving in the same way as x. E.g., consider the set SAT. Given a formula F, trivially satisfiable formulas G_y can be found for each $y \in \Sigma^*$. Then F is satisfiable if and only if $(F \wedge G_y)$ is satisfiable.

Berman and Hartmanis (1977) show that all "known" NP-complete sets are invertible paddable using similar ideas as above. This result becomes interesting because they also show that an NP-complete set is invertible paddable if and only if it is polynomially isomorphic to SAT. That is, the Berman-Hartmanis conjecture is true if and only if each NP-complete set is invertibly paddable. Additional results concerning such padding functions have been obtained recently by Young (1983). Especially, Young gives some examples of NP-complete sets (the k-creative sets) not known to be paddable.

Next we show that intractable sets cannot be P-close if they are paddable.

<u>Theorem 4.9</u>. If a paddable set is P-close then it is in P.

<u>Proof</u>. Suppose A is a paddable and P-close set. Let pad be the padding function for A according to Definition 4.8. Then for a polynomial q and each $x,y \in \Sigma^*$, $|pad(x, y)| \leq q(|x| + |y|)$. By the assumption that A is P-close there is a set $B \in P$ such that $A \Delta B$ is sparse. Hence, for a polynomial r and each $n \geq 0$, $|(A \Delta B)_{\leq n}| \leq r(n)$. Then choose a polynomial s such that for each $n \geq 0$,

$$s(n)/2 > r(q(n + \log s(n))).$$

(Choose s of higher degree than $r \circ q$).

The following algorithm computes A in polynomial time (letting $y_1, y_2, \ldots$ be the standard enumeration of Σ^*. Then, $|y_i| \leq \log i$.)

On input x, $|x| = n$,
compute $w_1 = pad(x, y_1), \ldots, w_{s(n)} = pad(x, y_{s(n)})$,
and accept the input x if and only if for more than
half of the w_i, $w_i \in B$.

This algorithm obviously runs in polynomial time because s is a polynomial. Further, the algorithm correctly computes A because the number of w_i for which $w_i \in B$ iff $x \notin A$ is at most

$$\begin{aligned} r(|pad(x, y_{s(n)})|) &\leq r(q(n + |y_{s(n)}|)) \\ &\leq r(q(n + \log s(n))) \\ &< s(n)/2 \quad \text{by the choice of } s. \end{aligned}$$

This proves that $A \in P$. □

Corollary 4.10. If each set in NP is P-close then P = NP.

Hence we have shown that the assumption that every set in NP has efficient and "almost correct" decision procedures leads to the unlikely consequence that P = NP. Observe that we do not have the result that the P-closeness of some NP-complete set already implies P = NP. This is still open. But for EXPTIME-complete sets such a strong result can be established. Notice that the following theorem improves a result by Yesha (1983) considerably. Yesha shows that no EXPTIME-complete set can be O(log log n)-close to a set in P, that is, there is no EXPTIME-complete set A and no set B in P such that $|(A \Delta B)_{=n}| = O(\log \log n)$. We lift Yesha's O(log log n)-bound to O(p(n)) where p is any polynomial.

Theorem 4.11. No EXPTIME-hard set is P-close. That is, no EXPTIME-hard set can form a sparse symmetric difference with a set in P.

The Proof uses the following lemma which is a stronger version of a result in (Balcázar und Schöning, 1985), and also in (Berman and Hartmanis, 1977).

Lemma 4.12. There exists a set A in EXPTIME such that

(i) every $\leq_m^P$-reduction f from A to some other set is one-one almost everywhere, i.e. there are at most finitely many pairs (x, y) with $x \neq y$ and $f(x) = f(y)$,

(ii) for every $D \in P$, $|(A \Delta D)_{=n}| = \Omega(2^n/n)$.

First we show how the lemma can be used to prove the theorem. Suppose B

s an EXPTIME-hard set, and for some $C \in P$, some polynomial p and ach $n \geq 0$, $|(B \Delta C)_{\leq n}| \leq p(n)$. By the fact that the set A from emma 4.12 is in EXPTIME and B is EXPTIME-hard, there is a $\leq_m^P$-eduction f reducing A to B. By Lemma 4.12 (i), f is one-one lmost everywhere. Let d be the number of finitely many exceptions in he one-one-ness of f. Let q be a polynomial bounding the run time or the computation of f. Then for each $x \in \Sigma^*$, $|f(x)| \leq q(|x|)$. efine the set $D = \{ x \mid f(x) \in C \}$. It is obvious that $D \in P$ since $\in P$, and f can be computed in polynomial time. Now we show that $A \Delta D$ s sparse which contradicts Lemma 4.12 (ii). Then the theorem is proved. or each $n \geq 0$ we have

$$
\begin{aligned}
|(A \Delta D)_{\leq n}| &\leq |(B \Delta C)_{\leq q(n)}| + d \\
&\leq p(q(n)) + d. \quad \square
\end{aligned}
$$

roof of Lemma 4.12. Let $P_1, P_2, \ldots$ be an enumeration of the olynomial time bounded Turing machine acceptors, and let $T_1, T_2, \ldots$ e an enumeration of the polynomial time bounded transducers (i.e., uring machines that compute functions). We also use the notation T_i or the function computed by T_i. Now the construction of the desired et A proceeds in stages such that, after stage n, membership of all trings of size n to either A or $\overline{A}$ is defined. In the construction, is a list of indices (of Turing machines) which still have to be onsidered in the stage-by-stage diagonalization process.

<u>stage 0</u>:
  $A := \emptyset$;
  $R := \emptyset$;

<u>stage $n > 0$</u>:
(1) <u>if</u> T_n on inputs of size n runs for at most 2^n steps
    <u>then</u> $R := R \cup \{n\}$;
  <u>for</u> $j \in R$ <u>do</u>
    <u>for</u> each y of size n <u>do</u>
      <u>for</u> each x, $x < y$ <u>do</u>
        <u>if</u> $T_j(x) = T_j(y)$ <u>then</u>
(2)   <u>begin</u>

```
              if  x ∉ A  then  A := A ∪ {y} ;
              R := R - {j} ;
              goto  M ;
           end ;
M:     Partition the  2^n  strings of size  n  into  n  sets  S_1,...,S_n
       (according to lexicographical order) such that
       |S_i| = [2^n/n],   1 ≤ i < n.
       for  k := 1  to  n  do
(3)       if  P_k  on inputs of size  n  runs for at most  2^n  steps  then
             for  z ∈ S_k  (except  z=x  and  z=y)  do
                if  z ∉ L(P_k)  then  A := A ∪ {z} ;
     end of stage n.
```

Let f be a function computable in polynomial time. Then there exist infinitely many indices $j_1, j_2, \ldots$ such that each of the transducers $T_{j1}, T_{j2}, \ldots$ computes f. By the fact that the function 2^n eventually majorizes every polynomial there must be a (smallest) $n \in N$ such that the index j_n in line (1) of the above construction enters the list R. This happens in stage j_n. Suppose the function f is not one-one almost everywhere, that is, there are infinitely many pairs (x, y) with $x \neq y$ and $f(x) = f(y)$. Since there are only finitely many smaller indices than j_n which, at that moment, might be contained in R, there must be a stage $m \geq j_n$ such that, in stage m, the compound statement starting in line (2) is entered with $j = j_n$. But then, $y \in? A$ is defined in such a way that f cannot be a $\leq_m^P$-reduction from A to any other set because $x \in A$ iff $y \notin A$, but on the other hand, $f(x) = f(y)$. Hence we have shown that every $\leq_m^P$-reduction from A to another set must be one-one almost everywhere.

Next we observe that $A \in$ EXPTIME. For deciding whether $x \in A$, only the stages $0,1,\ldots,|x|$ have to be performed, and each of these stages needs time $2^{O(|x|)}$. This is garanteed by the tests in line (1) and (3).

Finally, the second part of the construction starting at line M (does not disturb the first, and) garantees that for each k and almost every n

$$|(A \,\Delta\, L(P_k))_{=n}| \;\geq\; [2^n/n] - 2 \;.$$

This implies that for each set $D \in P$, $|(A \,\Delta\, D)_{=n}| = \Omega(2^n/n)$. □

4.3. "ALMOST FAST" ALGORITHMS AND COMPLEXITY CORES

A different kind of approximation algorithms was introduced by Meyer and Paterson (1979). They studied algorithms that are always correct but may not run in polynomial time on a sparse set of instances.

<u>Definition 4.13</u>. A set A is in the class APT (almost polynomial time) if there is a deterministic Turing machine M that accepts A, a polynomial p, and a sparse set S such that M runs for at most $p(|x|)$ steps for all $x \in \Sigma^* - S$.

It is easy to see that each set in APT has polynomial size circuits, moreover, each set in APT is P-close. An "almost polynomial time" machine according to Definition 4.13 can be equipped with a clock. If the allowed run time is over, and the machine has not yet accepted or rejected, then the new machine just rejects (or accepts). Then the new machine makes "mistakes" on at most a sparse set.

Further, each recursive tally set is in APT.

The question whether NP-complete sets can be in APT was studied by Meyer and Paterson (1979), and (implicitely) by P.Berman (1978). The results state that this is not possible unless P = NP. Using Mahaney's more recent result (Theorem 4.2), this is not hard to see: Suppose there is an NP-complete set A in APT (via machine M, polynomial p, and sparse set S according to Definition 4.13), and let a be a fixed element in A. Then the following function $\leq_m^P$-reduces A to the sparse set $(A \cap S) \cup \{a\}$ which implies P = NP by Theorem 4.2.

$$f(x) = \begin{cases} a, & \text{if } M \text{ on } x \text{ accepts within } p(|x|) \text{ steps,} \\ x, & \text{otherwise.} \end{cases}$$

These reflections show that APT is too strong a notion to allow the possibility of approximating NP-complete sets. But, nevertheless, in the following it is shown that the class APT can be used to establish an interesting connection to the notion of "complexity cores".

Definition 4.14. (Lynch, 1975) A set X is a (polynomial) complexity core for a set A if for each Turing machine M that accepts A, and each polynomial p, M runs for more than $p(|x|)$ steps on almost every input $x \in X$.

The notion of a complexity core is aiming to the question what (collections of) instances make an intractable problem difficult. Observe that the "natural" intractable problems always possess infinite collections of input instances for which the problem is easily decidable. E.g., the satisfiability of Boolean formulas is easily decidable for formulas in Horn form, or in conjunctive normal form with at most two literals per clause. A complexity core for SAT would be a collection of formulas that is uniformly hard, independent of the algorithm used to decide satisfiability.

A question that occurs naturally is whether such infinite collections of uniformly hard instances can always exist. This is answered by the following theorem.

Theorem 4.15. (Lynch, 1975) Each recursive set not in P has an infinite recursive complexity core.

The conditions necessary for a complexity class to permit such a core theorem are studied in (Even, Selman and Yacobi, 1985). An unsatisfying point about Theorem 4.15 is that there is no assertion about the density of the hard instances in the core set. E.g., for NP-complete sets it is interesting to know "how dense" complexity cores can be. As a simple measure for the density of a set we use the categories sparse/non-sparse.

Theorem 4.16. (Orponen and Schöning, 1984) A recursive set is in APT if and only if each of its complexity cores is sparse.

Proof. The foreward direction is trivial since each complexity core of a set in APT has to be (almost everywhere) included in the sparse set S from Definition 4.13.

Conversely, suppose A is a recursive set not being in APT. We will show that A has a non-sparse complexity core. Let $M_1, M_2, \ldots$ be an enumeration of all Turing machines that accept A (this enumeration is, in general, non-effective). Further, let $p_k(n) = n^k + k$, $k = 1,2,\ldots$ be an enumeration of polynomials. This "standard" enumeration has the following two properties that will be used in the following.

(i) For each n and k, $p_k(n) < p_{k+1}(n)$.

(ii) For each polynomial q there is an index k such that for all n, $q(n) \leq p_k(n)$.

The core set X is constructed in stages as follows.

<u>stage 0</u>:

$X_0 := \emptyset$;

$n_0 := 0$;

<u>stage k+1</u>:

If there is a finite set Y and a number $n > n_k$ satisfying the condition C_k below, then choose the smallest such pair (Y, n) (according to some total order) and let

$X_{k+1} := X_k \cup Y$;

$n_{k+1} := n$.

Here, C_k is the following condition:

$Y \subseteq \Sigma^n$, $|Y| > p_k(n)$, and for each $y \in Y$ and each $i \leq k$, M_i on input y runs for more than $p_k(n)$ steps.

Finally, let $X = \bigcup_k X_k$. We claim that condition C_k can be satisfied in each stage k, and thus, X becomes an infinite set. Suppose there is a stage $k+1$ where C_k is not satisfied for any (Y, n). This means that the set

$$H = \{ x \mid \text{for each } i \leq k,\ M_i \text{ on input } x \text{ runs for more than } p_k(|x|) \text{ steps} \}$$

is sparse (via polynomial p_k). Now consider the following Turing machine M: M has $n_1+\ldots+n_k$ tapes where n_i, $i \leq k$, is the number of tapes of M_i. M, on input x, simulates the behavior of $M_1,\ldots,M_k$ in parallel until the first of these machines accepts or rejects, and so does M. Clearly, $L(M) = A$ and

$H = \{ x \mid M$ on input x runs for more than $p_k(|x|)$ steps $\}$.

By the fact that H is sparse, it follows that $A \in APT$ (via M) contradicting the assumption that $A \notin APT$. Thus, X is an infinite set moreover, by the "thickness" of the added sets Y in each stage, the function $|X_{=n}|$ majorizes every polynomial (infinitely often). That is, X is non-sparse.

Finally, we have to show that X is a complexity core for A. Let M be a Turing machine that accepts A, and let q be a polynomial. Then there are indices i, j such that $M = M_i$ and $q \leq p_j < p_k$ (by clauses (i) and (ii) above). Then, by the definition of X, M runs for more than $q(|x|)$ steps on all inputs $x \in X$ with $|x| > \max\{i,j\}$. Thus, X is a complexity core for A. □

Observe that the core set X constructed above in general is not recursive since its construction is based on the non-effective enumeration of the machines M_i. But, in (Orponen and Schöning, 1985) it is shown that this construction (and the one from Theorem 4.15) can be made effective (but more involved), even such that the core set is in EXPTIME.

Further, by the fact that there are sparse sets having Σ^* as complexity core (i.e. "bi-immune" sets in the terminology of (Balcázar and Schöning, 1985)), it follows from Theorem 4.16 that there are P-close sets not being in APT.

We continue with two special applications of Theorem 4.16.

<u>Corollary 4.17</u>. If $P \neq NP$, then every NP-complete set has a non-sparse complexity core.

<u>Proof</u>. Using the fact that NP-complete sets cannot be in APT unless $P = NP$ (cf. Theorem 4.2 and the remarks after Defition 4.14), the result follows directly from Theorem 4.16. □

<u>Corollary 4.18</u>. Each paddable set not in P has a non-sparse complexity core.

<u>Proof</u>. This follows from Theorem 4.9, Theorem 4.16 and the observation that APT-sets are P-close. □

It is interesting to note here that Corollary 4.17 does not seem to be a direct consequence of Corollary 4.18 because the Berman-Hartmanis conjecture is still unsolved. In contrast to Corollary 4.18 we show that invertibly paddable sets cannot have too "fat" complexity cores.

Theorem 4.19. (Orponen and Schöning, 1984) No invertibly paddable set can have a complexity core that is co-sparse.

Proof. Let A be an invertible paddable set where pad and decode are the functions from Definition 4.8. Fix any string $x \in \Sigma^*$ and define $D_x = \{ z \mid z = pad(x, y) \text{ for some } y \in \Sigma^* \}$. It is obvious that $D_x \in P$ since $z \in D_x$ iff $pad(x, decode(z)) = z$. Further, for each $x \in A$, $D_x \subseteq A$, and for each $x \in \bar{A}$, $D_x \subseteq \bar{A}$. We will show that $|(D_x)_{\leq n}| = \Omega(2^{n^e})$ for some $e > 0$. From this, the theorem follows because the existence of such a set $D_x \in P$ within A (or, within $\bar{A}$) implies that there is a Turing machine that accepts A and runs in polynomial time on D_x. Thus, no complexity core of A can be co-sparse, otherwise it incudes parts of D_x.

Let q be a polynomial bounding the computation time for pad. Then for all x, y, $|pad(x, y)| \leq q(|x|+|y|)$. Since pad is one-one, for each $n \geq 0$ and $x \in \Sigma^*$,

$$|(D_x)_{\leq q(|x|+n)}| \geq |(\Sigma^*)_{\leq n}| = 2^{n+1} - 1 .$$

It follows that there is a constant e (essentially the reciprocal value of the degree of q) such that $|(D_x)_{\leq n}| = \Omega(2^{n^e})$. □

For EXPTIME-hard sets the assertions about the density of their complexity cores can be even improved.

Theorem 4.20. (Orponen and Schöning, 1984) For each EXPTIME-hard set B there is a constant e such that B has a complexity core X of density $\Omega(2^{n^e})$.

Proof. By Lemma 4.12, there is a set A in EXPTIME such that every $\leq_m^P$-reducion from A to any other set is one-one almost everywhere. Ko and Moore (1981) observed that such a language has Σ^* has its complexity core (that is, A is "bi-immune", cf. (Balcázar and Schöning, 1985)). This can be seen as follows. If M is a Turing machine that accepts A, and if p is a polynomial such that the set $C = \{ x \mid M$ on input x halts within $p(|x|)$ steps $\}$ is infinite, then C, $C \cap A$, and $C \cap \bar{A}$ are in P. At least one of the sets $C \cap A$ or $C \cap \bar{A}$ is infinite. W.l.o.g. this is $C \cap A$. Let y be a fixed element in $C \cap A$. Then the following function

$$f(x) = \begin{cases} y, & \text{if } x \in C \cap A, \\ x, & \text{otherwise.} \end{cases}$$

is computable in polynomial time and provides a $\leq_m^P$-reduction from A to A which is not one-one almost everywhere, contradicting the properties of A. Thus, Σ^* is a complexity core of A.

By the fact that B is EXPTIME-hard, there is a $\leq_m^P$-reduction g from A to B which, by Lemma 4.12, is one-one almost everywhere. Let q be a polynomial that bounds the computation time for g, and let d be the number of exceptions in the one-one-ness of f. Since Σ^* is a complexity core for A, $X = f(\Sigma^*)$ is a complexity core of B, and for the density of X follows

$$|X_{\leq q(n)}| \geq |(\Sigma^*)_{\leq n}| - d = 2^{n+1} - 1 - d.$$

Thus, $|X_{\leq n}| = \Omega(2^{n^e})$ for a suitable constant $e > 0$. □

4.4 RELATIVIZATION, SPARSE ORACLES, AND CIRCUITS

Meyer (stated in (Berman and Hartmanis, 1977)) observes that the class of sets having polynomial size circuits is exactly the class of sets recognizable in polynomial time relative to a sparse oracle. Intuitively, a set has polynomial size circuits if it can be computed efficiently with a "small" amount of additional side-information (very similar to the notion P/Poly). In the following we extend Meyer's result to all Σ-classes in the polynomial hierarchy, moreover, it is shown to be true for tally sets, too.

Theorem 4.21. For each $k \geq 0$, $\Sigma_k^P/\mathrm{Poly} = \bigcup\{ \Sigma_k^P(T) \mid T \subseteq \{0\}^* \} = \bigcup\{ \Sigma_k^P(S) \mid S$ is sparse $\}$.

Proof. Let A be in Σ_k^P/Poly for some $k \geq 0$. That is, there is a set $B \in \Sigma_k^P$ and a function $h \in \mathrm{Poly}$ such that $A = \{ x \mid \langle x, h(|x|)\rangle \in B \}$. Suppose w.l.o.g that for all n, $|h(n)| = p(n)$ for some polynomial p. Define a tally set T as follows.

$$T = \{ 0^{\langle n,i\rangle} \mid \text{the i-th symbol of } h(n) \text{ is } 1 \}.$$

Now the following procedure witnesses that A can be $\leq_m^P$-reduced to B by means of a function f computable in polynomial time - relative to oracle T. Thus, $A \in \Sigma_k^P(T)$. This function f is computed by the following algorithm.

```
begin
  input  x ;
  w := λ ;
  for  i := 1  to  p(|x|)  do
    if  0^<|x|,i>  ∈ T  then  w := w1
                            else  w := w0 ;
  output  <x,w> ;
end.
```

To prove the theorem it now suffices to show that $\Sigma_k^P(S) \subseteq \Sigma_k^P/\mathrm{Poly}$ for every sparse set S. Let $A \in \Sigma_k^P(S)$ for a sparse set S. Then there is a polynomial p and a deterministic, polynomial time bounded oracle machine M such that

$$A = \{ x \mid (\exists y_1)_p (\forall y_2)_p \ldots (Q_k y_k)_p \langle x, y_1, \ldots, y_k\rangle \in L(M,S) \}.$$

Since M is polynomially time bounded, there is a polynomial q such that no oracle query of M on inputs of the form

$$\langle x, y_1, \ldots, y_k\rangle, \quad |x| = n, \quad |y_i| \leq p(n), \quad i = 1,2,\ldots,k$$

is longer than $q(n)$. Let r be a polynomial such that for each $n \geq 0$,

$|S_{<n}| \leq r(n)$. Define a function $h : N \to \Sigma^*$ by

$$h(n) = \langle w_1, \ldots, w_t \rangle, \; t \leq r(q(n)), \text{ where } S_{\leq q(n)} = \{w_1, \ldots, w_t\}.$$

Clearly, $h \in$ Poly. Furthermore,

$$A = \{ x \mid (\exists y_1)_p (\forall y_2)_p \ldots (Q_k y_k)_p \langle x, h(|x|), y_1, \ldots, y_k \rangle \in L(M') \}$$

where M' is a deterministic Turing machine which, on input $\langle x, h(|x|), y_1, \ldots, y_k \rangle$, behaves like M on input $\langle x, y_1, \ldots, y_k \rangle$. Oracle queries of M are simulated by checking whether the query string is among the strings encoded by $h(|x|)$. If it is, then the answer "yes" is simulated, otherwise "no". Clearly, $L(M') \in P$, and thus the following language

$$B = \{ \langle x, w \rangle \mid (\exists y_1)_p (\forall y_2)_p \ldots (Q_k y_k)_p \langle x, w, y_1, \ldots, y_k \rangle \in L(M') \}$$

is in Σ_k^p. By the characterization $A = \{ x \mid \langle x, h(|x|) \rangle \in B \}$, it follows that $A \in \Sigma_k^p/\text{Poly}$. □

Some corollaries of this result follow.

Corollary 4.22. A set A has polynomial size circuits if and only if there is a sparse set (or, a tally set) to which A is $\leq_T^p$-reducible.

Observe that the reduction is actually a polynomial-time truth-table reduction (which we will not further discuss here).

Corollary 4.23. A set A has polynomial size generators if and only if there is a sparse set (or, a tally set) S such that $A \in NP(S)$.

Corollary 4.24. (Long, 1985) There are sets A in EXPSPACE such that for each sparse set S, $A \notin NP(S)$.

Proof. Corollary 2.7 and 4.23. □

.5. SELF-REDUCIBILITY

The self-reducibility property of SAT has already been used in the roof of Theorem 3.6. Also, the proof of Mahaney's theorem (Theorem 4.2) ses the self-reducibility property extensively (cf. Mahaney, 1982). In he following we will give a formal definition which will be general nough for the applications to follow. But also more general notions have een studied in the literature (cf. Meyer and Paterson, 1979; Ko, 1983; elman, 1983). The first use of "self-reducibility" (in a somewhat ifferent sense) was made by Schnorr (1976). It has been called the method of recursive definition" in (Karp and Lipton, 1980) and "recursive tructure" in (Yap, 1983). The following treatment follows (Ko and chöning, 1985) and (Balcázar, Book and Schöning, 1984abc).

efinition 4.25. A set A is (polynomially) self-reducible if there is deterministic, polynomial time bounded oracle machine M such that

(i) A = L(M, A),

(ii) M on inputs of size n queries the oracles for strings of size at most n-1.

he definitions given in (Meyer and Paterson, 1979) and (Ko, 1983) are omewhat more general (and complicated), but the above definition uffices for our purposes.

It is easy to see that SAT (under a suitable encoding of Boolean ormulas) is self-reducible: the size of a formula $F(x_1,x_2,...,x_n)$ hould be longer than both $F(0,x_2,...,x_n)$ and $F(1,x_2,...,x_n)$. Then AT's self-reducibility follows from the observation that $F(x_1,...,x_n)$ s satisfiable if and only if $F(0,x_2,...,x_n)$ is satisfiable or $F(1,x_2,...,x_n)$ is satisfiable. All "natural" NP-complete sets have he self-reducibility property although it is not known whether all NP-omplete sets are self-reducible, a situation very similar to the uestion whether all NP-complete sets are paddable.

Also note that QBF (and similarly, QBF_k) is self-reducible. This ollows from the following equivalences:

$(\exists x)\ F(x)$ is true iff $F(0)$ is true or $F(1)$ is true,

$(\forall x)\ F(x)$ is true iff $F(0)$ is true and $F(1)$ is true.

Note that each self-reducible set is in PSPACE (cf. Ko, 1983) sinc the "self-reduction tree" induced by the self-reducibility of a language can be evaluated (e.g. using alfa-beta-pruning) reusing the same space f each level of the tree. Such an algorithm can evaluate a self-reducible set using polynomial space (and exponential time).

Definition 4.25 can be interpreted as A being a "fixed point" unde the "operator" M. In fact, these fixed points are uniquely determined.

<u>Lemma 4.26</u>. (Balcázar, Book and Schöning, 1984c) Let A be self-reducible via the oracle machine M. Let B be a set such that for some $n \geq 0$, $L(M, B)_{\leq n} = B_{\leq n}$. Then $A_{\leq n} = B_{\leq n}$.

<u>Proof</u> by induction on n. On inputs of size 0 (the empty string) M is not allowed to query the oracle. Hence, the decision of M on input λ does not depend on the oracle. Thus, $L(M, B_{\leq 0}) = B_{\leq 0}$ and $A = L(M, A)$ implies $A_{\leq 0} = B_{\leq 0}$.

Suppose for some $n \geq 0$, $L(M, B)_{\leq n+1} = B_{\leq n+1}$. By Definition 4.25, $L(M, B)_{\leq n+1} = L(M, B_{\leq n})_{\leq n+1}$. Using the induction hypothesis $A_{\leq n} = B_{\leq n}$, we get $B_{\leq n+1} = L(M, A_{\leq n})_{\leq n+1} = A_{\leq n+1}$. □

This is the key lemma for the following theorem. It is somewhat technical, but we will obtain a number of interesting corollaries from i

<u>Theorem 4.27</u>. (Balcázar, Book and Schöning, 1984c) Let A be a self-reducible set, and for some $k \geq 0$, $A \in \Sigma_k^p/\mathrm{Poly}$ (or equivalently, $A \in \Sigma_k^p(S)$ for a sparse or tally set S, cf. Theorem 4.21). Then $\Sigma_2^p(A) \subseteq \Sigma_{k+2}^p$.

<u>Proof</u>. Let M be the self-reducing machine for A from Definition 4.2 Since $A \in \Sigma_k^p/\mathrm{Poly}$, there is a set $B \in \Sigma_k^p$ and a polynomial p such that for each $n \geq 0$ a string w, $|w| \leq p(n)$, exists with

$$A_{\leq n} = \{ x \mid \langle x, w\rangle \in B \}_{\leq n} .$$

In the following we use the abbreviation B_w for $\{ x \mid \langle x, w\rangle \in B \}$.

Let L be in $\Sigma_k^p(A)$. We have to show that $L \in \Sigma_{k+2}^p$. By the quantifier characterization of the classes in the (relativized) polynomial hierarchy, there is a polynomial q and a deterministic, polynomial time bounded Turing machine M' such that

$$L = \{ x \mid (\exists y)_q (\forall z)_q \langle x,y,z\rangle \in L(M', A) \}.$$

Let r be a polynomial such that no oracle query of M' on inputs of the form

$$\langle x,y,z\rangle, \quad |x| = n, \quad |y|, |z| \le q(n)$$

is longer than $r(n)$. (r depends on the run time of M', q, and the way of encoding triples). Now, L can be characterized as follows.

$$L = \{ x \mid (\exists w)_{por} [(\forall u)_r \ (u \in B_w \text{ iff } u \in L(M, B_w)) \text{ and } (\exists y)_q (\forall z)_q \ \langle x,y,z\rangle \in L(M', B_w)] \}.$$

This characterization is correct because the formula

$$(\forall u)_r \ (u \in B_w \text{ iff } u \in L(M, B_w))$$

is true if and only if $(B_w)_{\le r(n)} = L(M, B_w)_{\le r(n)}$. Using Lemma 4.26, this is true if and only if $A_{\le r(n)} = (B_w)_{\le r(n)}$. Thus, in the characterization of L above, we can use B_w instead of A as oracle for the machine M'. Remember that M' does not query the oracle for strings longer than $r(n)$.

All quantifiers in this characterization of L are polynomially bounded. The predicate has the form

$$(\exists w) [(\forall u) \ R(u, w) \text{ and } (\exists y)(\forall z) \ S(x, y, z, w)]$$

where R, S are both predicates in $P(\Sigma_k^p) = \Delta_{k+1}^p$. Using elementary transformations (the Tarski-Kuratowski-algorithm), the above predicate can be brought into the form

$$(\exists w)(\exists y)(\forall u)(\forall z) [R(u, w) \text{ and } S(x, y, z, w)].$$

This immediately shows that $L \in \Sigma^p_{k+2}$. □

An interesting point about Theorem 4.27 is that we need not assume that A is in the polynomial hierarchy - but it follows from the conclusion of the theorem.

A simple consequence of the theorem is Karp and Lipton's result (Theorem 4.3) that $NP \subseteq P/Poly$ implies $PH = \Sigma^p_2$. Choose $k = 0$ and $A = SAT$ in Theorem 4.25, then it follows that $\Sigma^p_3 = \Sigma^p_2(SAT) = \Sigma^p_2$, i.e., the polynomial hierarchy collapses to Σ^p_2. Another simple consequence is a recent result by Yap (1983).

<u>Corollary 4.28</u>. $PH/Poly = \Sigma^p_k/Poly$ implies $PH = \Sigma^p_{k+2}$.

<u>Proof</u>. The assumption $PH/Poly = \Sigma^p_k/Poly$ implies

$$QBF_{k+1} \in \Sigma^p_{k+1} \subseteq PH \subseteq PH/Poly \subseteq \Sigma^p_k/Poly.$$

Thus, using Theorem 4.27 (with $k+1$ and QBF_{k+1}),

$$\Sigma^p_{k+3} = \Sigma^p_2(QBF_{k+1}) = \Sigma^p_{k+2}.$$

This is equivalent to $PH = \Sigma^p_{k+2}$. □

The following corollary is a generalization of a result in (Karp and Lipton, 1980).

<u>Corollary 4.29</u>. $PSPACE \subseteq \Sigma^p_k/Poly$ implies $PSPACE = PH = \Sigma^p_{k+2}$.

<u>Proof</u>. Apply Theorem 4.27 with $A = QBF$. □

Some other consequences of Theorem 4.27 will be obtained in the next chapters.

CHAPTER 5

THE LOW AND HIGH HIERARCHIES

A low and a high hierarchy within the class NP have been defined and investigated in (Schöning, 1983). The definition is a translation of a recursion theoretic notion (cf. Lerman, 1983, pp. 75) into the context of polynomially time bounded computations. In a sense, low sets behave similar to P-sets, they have low "information content" (when used as an oracle), whereas the high sets contain the NP-complete sets, i.e. sets of high "information content". The classification of sets into those that are "low" and those that are "high" (definitions are given below) does not coincide with the classical complexity-theoretic hierarchies. Nevertheless, there exist certain relationships which will be explored in this chapter.

The original intention was to define more general versions of "NP-completeness" using the high hierarchy. But later, additionally, a connection between the low hierarchy and the polynomial size circuit sets could be established (cf. Ko and Schöning, 1985) which allows to view the polynomial hierarchy collapsing results (cf. Theorems 4.2-4.4, Corollaries 4.28, 4.29) in a new light.

In the following sections we develop the theory of the low and high hierarchies, we introduce a generalization from (Balcázar, Book and Schöning), and demonstrate a connection between lowness and polynomial size circuits.

5.1. DEFINITIONS AND ELEMENTARY RESULTS

For each set A in NP and $k \geq 1$, the following inclusions hold:

$$\Sigma_k^p \subseteq \Sigma_k^p(A) \subseteq \Sigma_{k+1}^p .$$

If $A \in P$, then the left inclusion becomes equality, and if A is NP-complete, the right inclusion becomes equality. The idea now is to measure the "lowness" (= similarity with P-sets) of a set in terms of the smallest k such that the left inclusion becomes equality. Analogously, we measure the "highness" (= similarity with NP-complete sets) in terms of the smallest k such that the right inclusion becomes equality. Now it is easy to see that a set cannot be simultaneously low and high unless the polynomial-time hierarchy collapses. That is, the following definitions of lowness and highness become meaningful under the assumption that the polynomial hierarchy is infinite.

Further, this idea will be generalized so that it is applicable to all sets, not only those in NP. The following definitions are direct translations of corresponding recursion-theoretic definitions (cf. Lerman 1983, pp. 75). The low and high hierarchies within the class NP have been introduced (in a slightly different way) in (Schöning, 1983). The extended low/high hierarchies are from (Balcázar, Book and Schöning, 1984ab).

Definition 5.1. (i) For each $k \geq 0$, $L_k^P = \{ A \in NP \mid \Sigma_k^P(A) \subseteq \Sigma_k^P \}$ is the low hierarchy in NP, and $H_k^P = \{ A \in NP \mid \Sigma_{k+1}^P \subseteq \Sigma_k^P(A) \}$ is the high hierarchy in NP.

(ii) For each $k \geq 1$, $EL_k^P = \{ A \mid \Sigma_k^P(A) \subseteq \Sigma_{k-1}^P(A \oplus SAT) \}$ is the extended low hierarchy, and for each $k \geq 0$, $EH_k^P = \{ A \mid \Sigma_k^P(A \oplus SAT) \subseteq \Sigma_k^P(A) \}$ the extended high hierarchy.

Note that $EL_k^P \cap NP = L_k^P$ $(k \geq 2)$ and $EH_k^P \cap NP = H_k^P$ $(k \geq 0)$. Further, it is obvious that $L_k^P \subseteq L_{k+1}^P$, $H_k^P \subseteq H_{k+1}^P$, $EL_k^P \subseteq EL_{k+1}^P$, and $EH_k^P \subseteq EH_{k+1}^P$. None of these inclusions is known to be proper. As noted above, the low and high hierarchies are disjoint if and only if the polynomial hierarchy is infinite.

Theorem 5.2 (Schöning, 1983) For each $k \geq 0$,

(i) if $PH \neq \Sigma_k^P$, then $L_k^P \cap H_k^P = \emptyset$,

(ii) if $PH = \Sigma_k^P$, then $L_k^P = H_k^P = NP$.

Proof. (i) If $A \in L_k^P \cap H_k^P$, then both $\Sigma_k^P(A) \subseteq \Sigma_k^P$ and $\Sigma_{k+1}^P \subseteq \Sigma_k^P(A)$. Thus it follows that $\Sigma_k^P = \Sigma_{k+1}^P$, hence $PH = \Sigma_k^P$.

(ii) The case $k = 0$ is obvious. Let $k \geq 1$ and $A \in NP$. Then, $\Sigma_k^P \subseteq \Sigma_k^P(A) \subseteq \Sigma_{k+1}^P$. From the assumption $PH = \Sigma_k^P$ it follows $\Sigma_k^P = \Sigma_k^P(A) = \Sigma_{k+1}^P$. Thus, $A \in L_k^P$ and $A \in H_k^P$. □

By "relativizing" this proof, it is easy to see that $A \in EL_k^P \cap EH_k^P$ if an only if $PH(A \oplus SAT) = \Sigma_{k-1}^P(A \oplus SAT)$. Hence, it follows that the extended low and high hierarcies are not disjoint, because there exist sets - like QBF - with $PH(QBF \oplus SAT) = \Sigma_{k-1}^P(QBF \oplus SAT) = PSPACE$ for each $k \geq 1$. Thus, $QBF \in EL_k^P \cap EH_k^P$ for all $k \geq 1$.

The bottom stages of these low and high hierarchies can be characterized in terms of certain polynomial time reducibities. For this purpose we define the strong nondeterministic Turing reduction $\leq_T^{sn}$ which was introduced and studied by Long (1982b).

$A \leq_T^{sn} B$ if and only if $A \in NP(B) \cap co\text{-}NP(B)$.

Strong nondeterministic Turing reducibility is a generalization of the more familiar γ-reducibility (Adleman and Manders, 1977; cf. Garey and Johnson, 1979). (γ-reducibility is nothing else than strong nondeterministic many-one reducibility). There are some languages known to be "γ-complete" (hence, $\leq_T^{sn}$-complete for NP) but not known to $\leq_T^P$-complete for NP. The following characterization of $\leq_T^{sn}$ is due to Selman (1978) (see also Schöning, 1983; Long, 1985).

Lemma 5.3. $A \leq_T^{sn} B$ if and only if $NP(A) \subseteq NP(B)$.

Proof. Let $A \in NP(B) \cap co\text{-}NP(B)$, and let M_1, M_2 be nondeterministic, polynomial time bounded oracle machines such that $A = L(M_1, B)$ and $\bar{A} = L(M_2, B)$. Let $L \in NP(A)$ be witnessed by a nondeterministic, polynomial time oracle machine M_3. Then $L \in NP(B)$ by means of the following machine constructed from M_1, M_2, and M_3:

On input x, simulate the behavior of M_3 on x. Each time M_3 queries its oracle for a string w, "guess" nondeterministically whether the answer is "yes" or "no", and then, verify the guess by running M_1 (or M_2, respectively) on input w. If M_1 (M_2) accepts, continue in the simulation of M_3, otherwise reject.

Conversely, suppose $NP(A) \subseteq NP(B)$. By the fact that $A \in NP(A)$ and $\bar{A} \in NP(A)$, it follows that $A \in NP(B) \cap co\text{-}NP(B)$. Thus, $A \leq_T^{sn} B$. □

Using this lemma the following theorem is immediate.

<u>Theorem 5.4</u>. (i) $L_0^P = P$,

(ii) $L_1^P = NP \cap co\text{-}NP$,

(iii) $H_0^P = \{ A \mid A$ is $\leq_T^P$-complete for $NP \}$,

(iv) $EH_0^P = \{ A \mid A$ is $\leq_T^P$-hard for $NP \}$,

(v) $H_1^P = \{ A \mid A$ is $\leq_T^{sn}$-complete for $NP \}$,

(vi) $EH_1^P = \{ A \mid A$ is $\leq_T^{sn}$-hard for $NP \}$.

<u>Proof</u>. (i),(iii),(iv) follow directly from the definition. (ii),(v) and (vi) follow from Lemma 5.3. □

The following diagram shows the low and high hierarchies in NP under the assumption that the polynomial hierarchy is infinite (cf. Theorem 5.2). Under this assumption, it is also known that there are sets in NP being neither in the low hierarchy nor in the high hierarchy (Schöning, 1983).

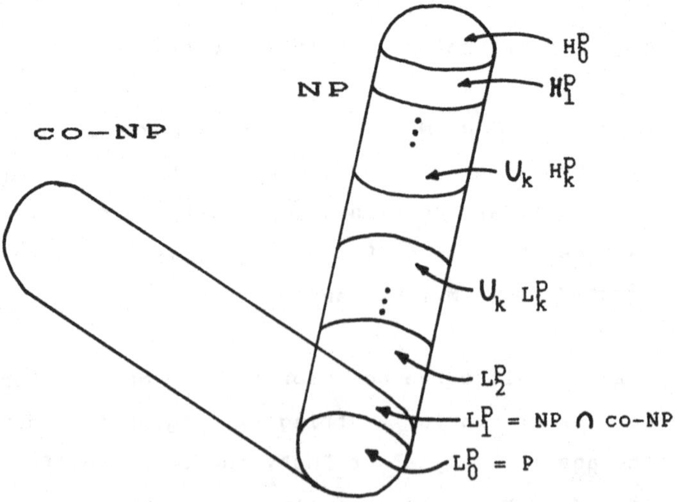

In chapter 4 we discussed several implications of the form

If some r-complete set for NP has some property E, then the polynomial hierarchy collapses to some level $n = n(r,E)$.

.g., we considered the cases where $r = \leq_m^P$ (or, $r = \leq_T^P$) and E is he property of being sparse (Theorems 4.2 and 4.3). That is, sparseness or the property of having polynomial size circuits) and NP-completeness eem to be incompatible. Very similar seems to be the situation with owness and highness:

If some set in H_i^P is also in L_j^P, then the polynomial hierarchy collapses to level $n = \max\{i,j\}$ (cf. Theorem 5.2).

e already showed the connection between NP-completeness and highness Theorem 5.4 (iii),(v)). In a sense, "highness" is a generalization of P-completeness. A natural question to ask now is whether lowness has omething to do with the properties E from above like being sparse or aving polynomial size circuits. In fact, such relationships will be hown to exist in the next chapter (e.g. $P/Poly \cap NP \subseteq L_3^P$).

.2. CIRCUIT-SIZE COMPLEXITY AND THE LOW HIERARCHY

One aim of this section is to show that sets having polynomial size ircuits are "low". Therefore, we have to develop a theory of "uniform-ess versus nonuniformness". In the following, a key definition is given. or each set A we introduce a class CIR(A) that may be interpreted as he collection of all polynomial size "circuits" which, in connection with ome "circuit interpreter", describe A.

All the material in this section is from (Ko and Schöning, 1985).

efinition 5.5. (i) For a set B and a string w let B_w be the et $B_w = \{ x \mid \langle x,w\rangle \in B \}$.

(ii) For each (multi-valued) function h from N to Σ^* let graph(h) = { $\langle 0^n,w\rangle$ | w is a value of h(n) }. h is called <u>total</u> if there is at least one value of h(n) for each $n \geq 0$, and h is called <u>polynomially bounded</u> if there is a polynomial p such that for each $n \geq 0$ and each value w of h(n), $|w| \leq p(n)$.

(iii) For each set A let CIR(A) be the following class of sets

$$\mathrm{CIR}(A) = \{ \mathrm{graph}(h) \mid h \text{ is total and polynomially bounded, and there exists a } B \in P \text{ such that for each } n \geq 0 \text{ and each value } w \text{ of } h(n),\ A_{\leq n} = (B_w)_{\leq n} \}.$$

Observe that a set A has polynomial size circuits (i.e., A ∈ P/Poly) if and onlf if CIR(A) is not the empty class. For a set A ∈ P/Poly, CIR(A) can contain nonrecursive sets. But, next we show that there exist such sets in the polynomial hierarchy relative to A.

<u>Lemma 5.6</u>. For each set A ∈ P/Poly, $\mathrm{CIR}(A) \cap \Pi_1^P(A) \neq \emptyset$.

<u>Proof</u>. By the fact that $\mathrm{CIR}(A) \neq \emptyset$ there is a set D = graph(h) ∈ CIR(A) where h is a total and polynomially bounded (possibly multi-valued) function. Let p be a polynomial such that $|h(n)| \leq p(n)$ for all $n \geq 0$, and let B be an appropriate set in P such that for each $n \geq 0$ and each value w of h(n), $A_{\leq n} = (B_w)_{\leq n}$.

Define a set C by

$$C = \{ \langle 0^n,w\rangle \mid n \geq 0,\ |w| \leq p(n),\ \text{and } A_{\leq n} = (B_w)_{\leq n} \}.$$

Clearly, $D \subseteq C$ and $C \in \mathrm{CIR}(A)$. C can be characterized as

$$C = \{ \langle 0^n,w\rangle \mid n \geq 0,\ |w| \leq p(n),\ \text{and } (\forall x, |x| \leq n)\,[\, x \in A \text{ iff } \langle x,w\rangle \in B \,] \}.$$

This characterization of C shows that $C \in \Pi_1^P(A)$. ◻

In contrast, the following lemma shows a reverse relationship between the polynomial hierarchy relativized to A, and the complexity of sets in CIR(A).

emma 5.7. (i) If $CIR(A) \cap \Sigma_k^P \neq \emptyset$, then $\Sigma_k^P(A) = \Sigma_k^P$.

(ii) If $CIR(A) \cap \Sigma_{k-1}^P(A) \neq \emptyset$, then $\Sigma_k^P(A) \subseteq \Sigma_{k-1}^P(A \oplus SAT)$, i.e, $A \in EL_k^P$.

roof. (i) Let L be a set in $\Sigma_k^P(A)$. Then, for a polynomial q and deterministic, polynomial time bounded oracle machine M,

$$L = \{ x \mid (\exists y_1)_q (\forall y_2)_q \ldots (Q_k y_k)_q \langle x, y_1, \ldots, y_k \rangle \in L(M, A) \}.$$

et $C = graph(h) \in CIR(A) \cap \Sigma_k^P$ where B is a suitable set in P (the circuit interpreter"). Suppose h is polynomially bounded via polyomial p. Let M′ be a Turing machine that, on an input of the form $x, y_1, \ldots, y_k, w\rangle$, operates like M on input $\langle x, y_1, \ldots, y_k\rangle$ where each racle query of M for some string z is simulated such that the answer yes" is assumed if and only if $\langle z, w\rangle \in B$. Clearly, M′ is polynomially ime bounded. Therefore, the set

$$D = \{ \langle x, w\rangle \mid (\exists y_1)_q (\forall y_2)_q \ldots (Q_k y_k)_q \langle x, y_1, \ldots, y_k, w\rangle \in L(M') \}$$

s in Σ_k^P. There is a polynomial r such that no oracle query of M n inputs of the form

$$\langle x, y_1, \ldots, y_k\rangle, \quad |x| = n, \quad |y_1|, \ldots, |y_k| \leq q(n),$$

an be longer than r(n). Now express L as follows.

$$L = \{ x \mid (\exists w)_{por} [\langle 0^{r(|x|)}, w\rangle \in C \text{ and } \langle x, w\rangle \in D] \}.$$

he correctness of this characterization of L follows from the fact hat $C \in CIR(A)$. Hence, L can be expressed by a predicate of the form $\exists(\Sigma_k^P$ and $\Sigma_k^P)$, that is, $L \in \Sigma_k^P$.

The proof of (ii) follows the proof of (i), but now the corresponding ets L, C, D are in $\Sigma_k^P(A)$, $CIR(A) \cap \Sigma_{k-1}^P(A)$, and Σ_k^P, respectively. ence, the final characterization of L has the form

$$\exists(\Sigma_{k-1}^P(A) \text{ and } \Sigma_k^P) = \exists(\Sigma_{k-1}^P(A) \text{ and } \Sigma_{k-1}^P(SAT)) \subseteq \Sigma_{k-1}^P(A \oplus SAT). \quad \square$$

Combining both lemmas, the following theorem is easily proved.

<u>Theorem 5.8</u>. (Balcázar, Book and Schöning, 1984ab) $P/Poly \subseteq EL_3^P$.

<u>Proof</u>. Let $A \in P/Poly$. By Lemma 5.6, there is a set $C \in CIR(A) \cap \Pi_1^P(A)$. Using Lemma 5.7 (ii) and the fact that $\Pi_1^P(A) \subseteq \Sigma_2^P(A)$, it follows that $A \in EL_3^P$. □

<u>Corollary 5.9</u>. All sets in NP having polynomial size circuits are in L_3^P. Therefore, no set in H_k^P can have polynomial size circuits unless the polynomial hierarchy collapses to $\Sigma_{max\{3,k\}}^P$. In particular, no $\leq_T^{sn}$-complete set for NP can have polynomial size circuits unless PH = Σ_3^P .

By Theorem 5.8, sets in P/Poly are extended low. Although there exist sets being both extended low and high, they cannot have polynomial size circuits (unless the polynomial hierarchy collapses).

<u>Theorem 5.10</u>. If $P/Poly \cap EH_k^P \neq \emptyset$, then $PH = \Sigma_{k+2}^P$.

<u>Proof</u>. Let $A \in P/Poly \cap EH_k^P$. Thus, $\Sigma_k^P(A \oplus SAT) \subseteq \Sigma_k^P(A)$. Using Theorem 4.21 (k=0), there is a sparse set S such that $A \in P(S)$. Hence,

$$\Sigma_{k+1}^P = \Sigma_k^P(SAT) \subseteq \Sigma_k^P(A \oplus SAT) \subseteq \Sigma_k^P(A) \subseteq \Sigma_k^P(S),$$

that is, $\Sigma_{k+1}^P \subseteq \Sigma_k^P/Poly$. Using an argument as in the proof of Corollary 4.28, it follows $PH = \Sigma_{k+2}^P$. □

5.3. LOWNESS OF SOME SPECIAL CLASSES

We showed that all sets having polynomial size circuits are included in the third level of the low hierarchy. For several subclasses of P/Pol such as the p-selective sets, the P-close sets, APT, R etc. this result can be improved (e.g. to the second level). The proofs will be similar t those in the previous chapter but, additionally, take advantage of the particular properties of these classes.

For the sets in P/Poly being self-reducible Theorem 5.8 can be strengthened.

Theorem 5.11. { A $\in$ P/Poly | A is self-reducible } $\subseteq EL_2^P$.

Proof. Follows from Theorem 4.27 (k=0). □

Corollary 5.12. Each self-reducible set in NP having polynomial size circuits is in L_2^P. Therefore, no set in H_k^P can be both self-reducible and in P/Poly unless PH = $\Sigma^P_{max\{2,k\}}$.

Notice that Theorem 4.3 is implied by Corollary 5.12. If NP $\subseteq$ P/Poly, then there is a self-reducible set (namely SAT) in the class P/Poly $\cap$ $H_0^P \cap L_2^P$.

For the p-selective sets, using the methods from chapter 2.5, a certain improvement of Lemma 5.6 is possible.

Lemma 5.13. If A is a p-selective set in Σ_k^P, then CIR(A) $\cap$ Δ^P_{k+1} $\neq \emptyset$.

Proof. Let f be a selector function for A according to Definition 2.15. By Lemma 2.16, for a polynomial p and each $n \geq 0$ there is a set $D_n \subseteq A_{\leq n}$, $|D_n| \leq p(n)$, such that

$$A_{\leq n} = \{ x \mid \text{for some } y \in D_n,\ f(x,y) = x \text{ or } f(y,x) = x \}_{\leq n} .$$

Define

$$B = \{ \langle x,w\rangle \mid w \text{ encodes a set of strings } \{y_1,\dots,y_k\} \text{ such that for some } i \leq k,\ f(x,y_i) = x \text{ or } f(y_i,x) = x \}.$$

Clearly, B $\in$ P. But then, for a suitable chosen polynomial q,

$$C = \{ \langle 0^n,w\rangle \mid |w| \leq q(n) \text{ and } w \text{ encodes a set of strings included in } A_{\leq n} \text{ such that } A_{\leq n} = (B_w)_{\leq n} \}$$

is in CIR(A). We claim that $C \in \Delta^P_{k+1}$. First observe that for each w encoding a subset of A, $\langle x,w\rangle \in B$ implies $x \in A$ (because f is a selector function for A). Using this observation, C can be characterized as follows.

$\langle 0^n,w\rangle \in C$ if and only if

$[|w| \leq q(n)]$ and [w encodes a subset of $A_{\leq n}$]

and $[(\forall x)_n\ (x \in A \text{ iff } x \in B_w)]$

if and only if

$[|w| \leq q(n)]$ and $[(\forall x)_n$ (x is contained in w $\rightarrow$ $x \in A)]$

and $[(\forall x)_n\ (x \in A \rightarrow x \in B_w)]$.

This shows that C can be expressed as $C = C_1 \cap C_2 \cap C_3$ where $C_1 \in P$, $C_2 \in \Sigma^P_k$, and $C_3 \in \Pi^P_k$. Thus, $C \in \Delta^P_{k+1}$. □

For p-selective sets we obtain the following strengthening of Corollary 5.9.

Theorem 5.14. $\{ A \in NP \mid A$ is p-selective $\} \subseteq L^P_2$. Therefore, no set in H^P_k can be p-selective unless $PH = \Sigma^P_{max\{2,k\}}$.

5.4. LOWNESS OF BPP

Especially interesting is the case of the class BPP. In section 3.[?] it was shown that BPP actually can be characterized very similar to the definition of P/Poly where, additionally, it is required that among the exponential number of "circuits" an overwhelming number of them correctly describes the set in question. It will be exactly this property which allows us to test (with a Σ^P_2-predicate) whether a "circuit" in CIR(A) correctly describes A. Intuitively, a "circuit" is OK if and only if behaves on A like a "hugh" number of other potential circuits. The ide[a] used in proving the following lemma are similar to those in (Lautemann, 1983) and (Zachos and Heller, 1984).

Lemma 5.15. If $A \in BPP$, then $CIR(A) \cap \Sigma_2^p \neq \emptyset$.

Proof. Let $A \in BPP$. Then by Theorem 3.10 there is a set $B \in P$ and a polynomial p such that among the $2^{p(n)}$ strings y of size $p(n)$ at least $(1 - 2^{-n})\cdot 2^{p(n)}$ of them satisfy

$$A_{=n} = \{ x \mid \langle x,y \rangle \in B \}_{=n} .$$

Define $C \in CIR(A)$ as

$$C = \{ \langle 0^n,y \rangle \mid |y| = p(n) \text{ and } A_{=n} = \{ x \mid \langle x,y \rangle \in B \}_{=n} \}.$$

We will show that $C \in \Sigma_2^p$. Loosely speaking, C will be characterized by a predicate stating

$$\langle 0^n,y \rangle \in C \text{ iff there are "many" } w,\ |w| = p(n), \text{ such that } (\forall x,\ |x| = n)\ [\ \langle x,w \rangle \in B \text{ iff } \langle x,y \rangle \in B\ ].$$

The notion of "many" used above should be such that it majorizes the quantity $2^{-n}.2^{p(n)} = 2^{p(n)-n}$ and is expressible by a Σ_2^p-predicate. The details are given in the following claim. Here, $\oplus$ denotes bitwise addition modulo 2.

Claim. Let E be a subset of $\Sigma^{p(n)}$.

(i) $(\exists\, u=\langle u_1,\dots,u_{p(n)} \rangle,\ |u_i|=p(n))\ (\forall\, v,\ |v|=p(n))$
$[$ for some $i \leq p(n)$, $u_i \oplus v \in E\ ]$
implies $|E| > 2^{p(n)-n}$.

(ii) $(\forall\, u=\langle u_1,\dots,u_{p(n)} \rangle,\ |u_i|=p(n))\ (\exists\, v,\ |v|=p(n))$
$[$ for all $i \leq p(n)$, $u_i \oplus v \notin E\ ]$
implies $|\Sigma^{p(n)} - E| > 2^{p(n)-n}$.

Using this claim, the theorem easily follows when defining E as

$$w \in E \text{ iff } (\forall x,\ |x| = n)\ [\ \langle x,w \rangle \in B \text{ iff } \langle x,y \rangle \in B\ ].$$

Proof of Claim. (i) Fix some $u = \langle u_1,\dots,u_{p(n)} \rangle$ such that for each v, $|v| = p(n)$, there is an $i \leq p(n)$ with $u_i \oplus v \in E$. Divide $\Sigma^{p(n)}$

into (not necessarily disjoint) subsets $V_1,\ldots,V_{p(n)}$ where $V_i = \{ v \in \Sigma^{p(n)} \mid u_i \oplus v \in E \}$. Clearly, for some $j \leq p(n)$, $|V_j| \geq 2^{p(n)}/p(n)$. Thus, $|E| \geq |V_j| \geq 2^{p(n)}/p(n) > 2^{p(n)-n}$.

(ii) Divide the set $U = \{ \langle u_1,\ldots,u_{p(n)}\rangle \mid |u_1| = \ldots = |u_{p(n)}| = p(n) \}$ into (not necessarily disjoint) subsets $U_1, \ldots, U_{2^{p(n)}}$ where $U_i = \{ \langle u_1,\ldots,u_{p(n)}\rangle \mid u_1 \oplus v_i \notin E, \ldots, u_{p(n)} \oplus v_i \notin E \}$. Here, v_i is the i-th string of size $p(n)$. Clearly, for some $j \leq 2^{p(n)}$, $|U_j| \geq |U|/2^{p(n)} = 2^{p^2(n)-p(n)}$. Then it follows $|\Sigma^{p(n)} - E|^{p(n)} \geq |U_j|$, thus $|\Sigma^{p(n)} - E| \geq 2^{p(n)-1} > 2^{p(n)-n}$. □

<u>Theorem 5.16</u>. (Zachos and Heller, 1984) $\Sigma_2^P(\mathrm{BPP}) = \Sigma_2^P$.

<u>Proof</u>. Lemma 5.15 and Lemma 5.7 (i). □

<u>Corollary 5.17</u>. (Sipser, 1983; Lautemann, 1983) $\mathrm{BPP} \subseteq \Sigma_2^P \cap \Pi_2^P$.

<u>Corollary 5.18</u>. The classes R and $\mathrm{BPP} \cap \mathrm{NP}$ are included in L_2^P. Therefore, no set in H_k^P can be in R or BPP unless $\mathrm{PH} = \Sigma^P_{\max\{k,2\}}$. In particular, no $\leq_T^{sn}$-complete set for NP can be in R or BPP unless $\mathrm{PH} = \Sigma_2^P$.

Note that Corollary 5.18 answers an open problem in (Adleman and Manders, 1979): Assuming $\mathrm{PH} = \Sigma_2^P$, there is no γ-complete set in R.

Consider the class R. From Corollary 5.15, $\mathrm{CIR}(A) \cap \Sigma_2^P \neq \emptyset$ for each set $A \in R$. Thus, $R \subseteq L_2^P$ (Corollary 5.18). But notice that in (Ko and Schöning, 1985) an even stronger statement is proved: For each $A \in R$, $\mathrm{CIR}(A) \cap \Pi_1^P \neq \emptyset$. Of course, this does not change the conclusion $R \subseteq L_2^P$.

5.5. REFINEMENT OF THE LOW AND HIGH HIERARCHIES

We introduce a certain refinement of the low and high hierarchies in NP based on the Δ-classes of the polynomial hierarchy instead of the Σ-classes. It is shown that some classes of sets connected with sparseness can be located in this refined low hierarchy.

Definition 5.19. For each $n \geq 1$, let $\hat{L}_n^P = \{ A \in NP \mid \Delta_n^P(A) \subseteq \Delta_n^P \}$ and $\hat{H}_n^P = \{ A \in NP \mid \Delta_{n+1}^P \subseteq \Delta_n^P(A) \}$.

Note that $\hat{L}_1^P = L_0^P = P$ and $\hat{H}_1^P = H_0^P = \{ A \mid A$ is $\leq_T^P$-complete for NP $\}$. The following theorem summarizes some elementary properties of these hierarchies.

Theorem 5.20. For each $k \geq 1$,

(i) $L_{k-1}^P \subseteq \hat{L}_k^P \subseteq L_k^P$,

(ii) $H_{k-1}^P \subseteq \hat{H}_k^P \subseteq H_k^P$,

(iii) $PH = \Delta_k^P$ implies $\hat{L}_k^P \cap \hat{H}_k^P = \emptyset$,

(iv) $PH = \Delta_k^P$ implies $\hat{L}_k^P = \hat{H}_k^P = NP$.

Proof. (i) and (ii) follow immediately from the definitions. The proof of (iii) and (iv) is similar to Theorem 5.2. □

It is not known whether the inclusions in Theorem 5.20 (i),(ii) are proper.

We obtained the results of the last chapters by considering the complexity of particular sets in the class CIR(A) where A is taken from a class having polynomial size circuits. Recall that CIR(A) is a collection of graphs of functions describing "circuits" for A. That is, we analyzed the checking problem for these functions. The checking problem for a function $h : N \to \Sigma^*$ means, given $\langle 0^n, w \rangle$, to determine whether w is a value of h(n). Now we want to actually evaluate the function h (now h is assumed to be single-valued). In general, the complexity of "checking" and "evaluating" can be quite different (cf. Valiant, 1976). Actually the whole theory of NP-completeness relies on this dichotomy: to check a satisfying assignment is easy, but to evaluate one might be hard.

First we prove a somewhat technical lemma that makes an assertion about the complexity of evaluating $0^n \mapsto h(n)$ in connection with sparse sets.

Lemma 5.21. Let S be a sparse set. Then the function $0^n \mapsto \langle S_{\leq n} \rangle$ is computable in polynomial time - relative to the oracle set prefix(S) = $\{ \langle 0^n, x \rangle \mid x$ is a prefix of some string of size at most n in $S \}$.

<u>Proof</u>. We have to show that an algorithm exists that computes $S_{\le n}$ in polynomial time, and uses prefix(S) as an oracle. The idea is to search through $\Sigma^0, \Sigma^1, \ldots, \Sigma^n$ systematically for prefixes of strings in $S_{\le n}$. The algorithm will be polynomial time bounded since S (and hence, prefix(S)) is sparse. Similar ideas have been used in (Mahaney, 1982; Long, 1982a; Long, 1985). We leave the verification to the reader.

```
begin
  input  0^n ;
  T := ∅ ;
  L := {λ} ;
  while  L ≠ ∅  do
    begin
      z := first element in  L ;
      L := L - {z} ;
      if  <0^|z|,z> ∈ prefix(S)  then  T := T ∪ {z} ;
      if  |z| < n  and  <0^n,z0> ∈ prefix(S)  then  L := L ∪ {z0} ;
      if  |z| < n  and  <0^n,z1> ∈ prefix(S)  then  L := L ∪ {z1} ;
    end ;  {while}
  output  <T> ;
end.
```

□

Lemma 5.21 can be used to show the lowness of several classes which use sparse sets in their respective definitions.

<u>Theorem 5.22</u>. (i) $\{ A \in NP \mid A$ is sparse $\} \subseteq \hat{L}_2^P$,

(ii) $APT \cap NP \subseteq \hat{L}_2^P$,

(iii) $\{ A \in NP \mid A$ is P-close $\} \subseteq \hat{L}_3^P$.

<u>Proof</u>. (i) Let A be a sparse set in NP. It suffices to show that $NP(A) \subseteq \Delta_2^P$, because $P(\Delta_2^P) = \Delta_2^P$. Hence, let $L \in NP(A)$. Let M be a nondeterministic, polynomial time bounded oracle machine with $L = L(M, A)$. Let $p(n)$ be a polynomial bounding the size of oracle queries

on inputs of size n. By the assumption that $A \in NP$, it follows that prefix(A) $\in$ NP. (On input $<0^n,x>$, guess nondeterministically a string y, $|xy| \leq n$, and verify that $xy \in A$). Using Lemma 5.21, $<A_{\leq n}>$ can be computed in polynomial time (from 0^n) - relativ to oracle prefix(A). Now L can be seen to be in Δ_2^P as follows. On input x, $|x| = n$, first compute (using prefix(A) as oracle) $w = <A_{\leq p(n)}>$, then ask the oracle for "$<x,w> \in? B$" where B is the following set.

$$B = \{ <x,w> \mid w \text{ encodes a set of strings } y_1,\ldots y_k \text{ and } x \in L(M, \{y_1,\ldots,y_k\}) \}.$$

Clearly, $B \in NP$. Thus, L can be computed by a deterministic, polynomial time bounded oracle machine using as oracles prefix(A) and B, both are sets in NP, hence it follows $L \in \Delta_2^P$.

(ii) Let $B \in APT \cap NP$. Then there is a deterministic Turing machine M, $L(M) = B$, a polynomial p, and a sparse set S (see Definition .13) such that M runs for at most $p(|x|)$ steps on all inputs from $\Sigma^* - S$. Define $S' \subseteq S$ as $S' = \{ x \mid M \text{ on input } x \text{ runs for at most } p(|x|) \text{ steps} \}$. Obviously, $S' \in P$, and hence $A := S' \cap B \in NP$. Since A contains all "hard" instances w.r.t the time bound p(n), A can be used as an oracle to compute B in polynomial time. Thus $B \in P(A)$, and $\Delta_2^P(B) \subseteq \Delta_2^P(A) \subseteq \Delta_2^P$ by part (i) of the theorem.

(iii) Suppose A is P-close and in NP. Then there is a set $C \in P$ such that $A \Delta C$ is sparse. We have $A \Delta C \in \Delta_2^P$ because $A \in NP$ and $C \in P$. Hence, prefix$(A \Delta C) \in \Sigma_2^P$. Let L be in $\Sigma_2^P(A)$. Then we have

$$L \in \Sigma_2^P(A) \subseteq \Sigma_2^P(P(A \Delta C)) = \Sigma_2^P(A \Delta C).$$

Then, similar to part (i), a deterministic, polynomial time bounded algorithm exists that computes L and operates relative to the oracles prefix$(A \Delta B) \in \Sigma_2^P$ and a similar set B now being in Σ_2^P. It follows $\Sigma_2^P(A) \subseteq P(\Sigma_2^P) = \Delta_3^P$, and thus $\Delta_3^P(A) \subseteq \Delta_3^P$. □

<u>Corollary 5.23</u>. No set in $\hat{H}_k^P$ can be sparse (or, in APT) unless PH $= \Delta_{max\{k,2\}}^P$. No set in $\hat{H}_k^P$ can be P-close unless PH $= \Delta_{max\{k,3\}}^P$.

From Theorem 5.22 (i) we obtain some separation results for some bottom levels of the low hierarchy under certain assumptions about the exponential time classes.

Corollary 5.24. (i) EXPTIME $\neq$ NEXPTIME implies $\hat{L}_1^P \neq \hat{L}_2^P$.

(ii) NEXPTIME $\neq$ co-NEXPTIME implies $L_1^P \neq \hat{L}_2^P$.

Proof. If EXPTIME $\neq$ NEXPTIME, then there are tally (hence sparse) sets in NP − P (cf. Theorem 4.6). Using Theorem 5.22, these sets are in $\hat{L}_2^P$ − P.

(ii) Very similar, under the assumption NEXPTIME $\neq$ co-NEXPTIME, there are tally sets in NP − (NP $\cap$ co-NP) (cf. Book, 1974). □

The diagram on the following page summarizes all known inclusions that hold within the class NP. Notice that the sparse sets and the co-sparse sets apparently do not behave in the same way in NP. Sparse sets are in $\hat{L}_2^P$ whereas co-sparse sets are not even known to be in L_2^P. A similar non-symmetry has been observed in a relativized context before. There are relativizations for which P $\neq$ NP, there are co-sparse sets in NP − P, but there are no sparse sets in NP − P (Hartmanis, Immerman and Sewelson, 1983).

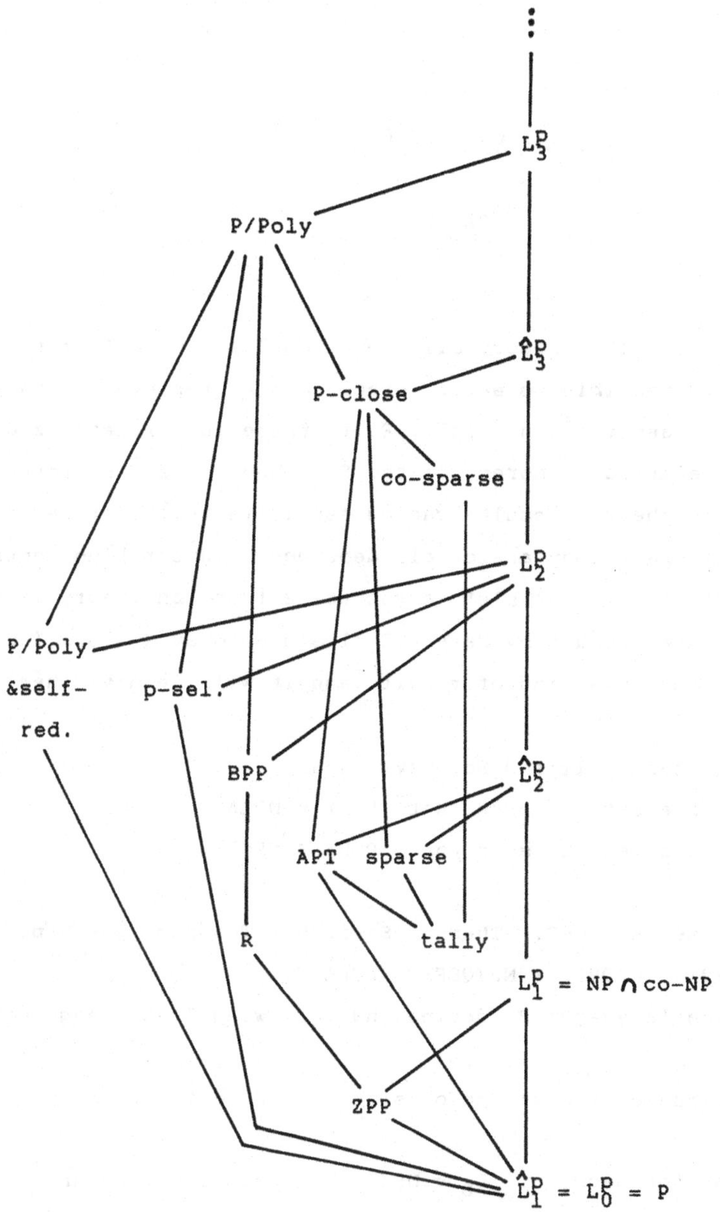

Inclusion structure of low sets in NP

CHAPTER 6

ORACLES

In recursive function theory all known results can be "relativized", i.e. it is always possible to add an oracle and to reprove the same theorem in the presence of an oracle. E.g., there is a relativized version of the recursion theorem, of the S_{mn}-theorem, and so forth.

In complexity theory, results can be sensitive to the presence of an oracle. In general, they are oracle-dependent. In striking contrast to the "relativization principle" in recursive function theory is the following well-known result by Baker, Gill and Solovay (1975). Its proof is very typical for this kind of result, and it will be given here.

Theorem 6.1. (Baker, Gill and Solovay, 1975)

(i) There is a set A such that $P(A) = NP(A)$.

(ii) There is a set B such that $P(B) \neq NP(B)$.

Proof. (i) Choose $A = QBF$. Then by Savitch's Theorem (see Hopcroft and Ullman, 1979), $P(QBF) = NP(QBF) = PSPACE$.

(ii) For each language B define the following "test-language"

$$L_B = \{ 0^n \mid \text{there is a string of size } n \text{ in } B \}.$$

It is clear that for each B, $L_B \in NP(B)$. (Guess a string and verify) Now we construct a particular set B such that $L_B \notin P(B)$. Then the theorem follows. B is constructed in stages. Let $M_1, M_2, \ldots$ be an enumeration of the deterministic, polynomial-time bounded oracle machines, and let $p_1, p_2, \ldots$ be an enumeration of their respective polynomial time bounds.

stage 0:

$B_0 := \emptyset$;

$n_0 := 0$;

<u>stage k > 0</u>:

    n_k := the smallest natural number $m > n_{k-1}$ such that

        (i) $m > \max\{\ p_i(n_{k-1})\ |\ i < k\ \}$,

        (ii) $2^m > p_k(m)$;

    <u>if</u> $0^{n_k} \in L(M_k, B_{k-1})$ <u>then</u> $B_k := B_{k-1}$

                             <u>else</u> $B_k := B_{k-1} \cup \{w\}$;

    (Here, w is the first string of size n_k in lexicographic order
that is <u>not</u> queried by M_k when operating on input 0^{n_k} and
using oracle B_{k-1}. Such a w always exists by clause (ii) above).

    Let $B = \bigcup_k B_k$. Suppose that $L_B \in P(B)$. Then there is an index
such that $L_B = L(M_k, B)$. By clause (i) in the above construction,
$0^{n_k} \in L(M_k, B)$ if and only if $0^{n_k} \in L(M_k, B_k)$. Moreover, by the choice
of w in stage k of the construction, $0^{n_k} \in L(M_k, B_k)$ if and only
if $0^{n_k} \in L(M_k, B_{k-1})$. Therefore, in stage k, B_k is defined in such
a way that

$$0^{n_k} \in L(M_k, B_{k-1}) \quad \text{iff} \quad \text{there is } \underline{\text{no}} \text{ string of size } n_k \text{ in } B_k.$$

Thus it follows that $0^{n_k} \in L(M_k, B) \,\Delta\, L_B$, contradicting the assumption
that $L(M_k, B) = L_B$. Hence $L_B \notin P(B)$, and thus $P(B) \neq NP(B)$. $\square$

    Observe that the set B constructed in the proof of Theorem 6.1 is
sparse because there is at most one string of each size in B. Moreover,
B could be made "arbitrarily sparse" by choosing the "diagonalization
points" n_k arbitrarily far apart.

    On the background of this result many authors suspected that "P = NP"
might be independent of the formal systems of reasoning in mathematics,
like number theory (cf. Hartmanis and Hopcroft, 1976). At least one can
say that new proof techniques seem to be necessary to solve the P-NP-
question (and other oracle dependent questions). That is, proof techniques
(not stemming from recursion theory) that do not carry over when intro-
ducing oracles.

    Another explanation for this "Baker-Gill-Solovay-phenomenon" is given
in (Book, Long and Selman, 1984). The diagonalization method in the proof
of Theorem 6.1 does not seem to have anything to do with the nature of
the question whether P = NP. Rather, the definition of "relativization"

seems to be to strong. A deterministic Turing machine, naturally, can ask only so many oracle questions as its time bound permits. But a nondeterministic oracle machine can do a number of queries that is exponential compared with its time bound. This very fact made the diagonalization proof of Theorem 6.1 possible. Observe that for nondeterministically recogizing the test-language L_B an exponential number of queries is necessary, and this allows to "fool" all deterministic machines. Hence, it seems reasonable to study relativations that are "fair", in the sense that the access of the nondeterministic machine to the oracle is restricted in the same way it is restricted for the deterministic machine naturally. Book, Long and Selman (1984) study several versions of polynomially restricting the access to the oracle of nondeterministic, polynomial time oracle machines, and the general pattern of result obtained is the following.

$$P = NP \quad \text{iff} \quad (\forall A) \quad P(A) = NP_b(A).$$

Here, $NP_b(\ )$ is a certain restriction of $NP(\ )$ where only a polynomial number of oracle queries is allowed (definition below). The restriction is such that $NP_b(\emptyset) = NP(\emptyset) = NP$. Hence, a potential proof of "$P = NP$" <u>does</u> relativize when using this restricted notion of relativization. The authors call a result like this a "positive relativization". We will also study a different type of positive relativization in section 6.1. The access to the oracle is not restricted, but the oracles are required to be sparse. Analogous results as the line above can be obtained.

A certain flaw in these positive relativization results is the fact that the main proof direction (from left to right in the above line) uses the assumption $P = NP$ which is believed to be false. One might object that everything can be proved from a false assumption. In section 6.2 we prove some dual results of the form

$$C \neq D \quad \text{iff} \quad (\forall \text{ sparse } A) \quad C(A) \neq D(A)$$

which we call a "negative relativization" (not quite in the sense of (Book, Long and Selman, 1984)).

In section 6.3, the question is studied how the polynomial hierarchy can exist with more than three distinct levels. Recall that the sets in

Σ_k^P are obtained as the sets in NP(B) where the oracle B ranges over the sets in Σ_{k-1}^P. Among others, it is shown that if $A \in \Sigma_k^P - \Delta_k^P$ $(k \geq 2)$, and $A \in NP(B)$ for a set B in Σ_{k-1}^P, then B cannot be sparse.

6.1. POSITIVE RELATIVIZATIONS

As discussed above, it is reasonable to restrict the access to the oracle of nondeterministic machines in a certain way. In the following we present one possibility to do this. It is the weakest restriction from (Book, Long and Selman, 1984) which still permits to prove a positive relativization result.

Definition 6.2. For each set A let $NP_b(A)$ be the class of all sets recognizable by some nondeterministic, polynomial time bounded oracle machine M relative to oracle A. Additionally it is required that there exists a polynomial p (depending on M) such that M on input x when using oracle A does not make more than $p(|x|)$ oracle queries in the computation tree induced by M, x, and A.

Notice that M is allowed to do an even exponential number of queries when using a different oracle set as A. That is, in the whole computation tree induced by M and x (ignoring the oracle, i.e. treating the oracle queries as nondeterministic branches) there might occur much more oracle queries than in the subtree that respects the answers to oracle A.

Theorem 6.3. (Book, Long and Selman, 1984)

P = NP if and only if for each oracle set A, $P(A) = NP_b(A)$.

Proof. The direction from right to left is trivial since $NP_b(\emptyset) = NP(\emptyset) = NP$.

Conversely, suppose P = NP and let L be a set in $NP_b(A)$. Then there exits an oracle machine M and a polynomial p according to

Definition 6.2. Suppose, w.l.o.g. that M always stops in a unique accepting configuration. (The reason for this constraint is that there is no exponential number of accepting configurations). Define the following set H.

H = { <x, C, D> | C is a configuration occuring in the computation tree of M on x, D is a prefix of some configuration C′ of M, such that C′ is either an accepting or a query configuration, and there is a computation path from C to C′ on which there is no other query configuration }.

It is easy to see that $H \in NP$, and by the assumption $P = NP$, $H \in P$. Observe that for each configuration C occuring in the computation tree induced by M, input x, <u>and oracle A</u>, the number of D′s with $\langle x, C, D\rangle \in H$ is polynomially bounded in $|x|$ (by the existence of the polynomial p and by the assumption above). Hence, using the "prefix searching technique" from Lemma 5.21, the function

$$f_H(\langle x, C\rangle) = \langle\{ C' \mid C' \text{ is a configuration of } M \text{ and } \langle x, C, C'\rangle \in H \}\rangle$$

is computable in polynomial time on the domain $\{ \langle x, C\rangle \mid C$ occurs in the computation tree of M, x, and A }. Using this function f_H (only on this domain) the following algorithm computes the set L in polynomial time relative to oracle set A, hence $L \in P(A)$.

```
begin
  input  x ;
  list := {C_M(x)} ;   {C_M(x) is the initial configuration of M on x}
  while  list ≠ ∅  do
    begin
      C := some element of list ;
      list := list - {C} ;
      compute  f_H(<x, C>),  and let the result be  <{C_1,...,C_m}>
      where  0 ≤ m ≤ p(|x|) ;
      for  i:= 1  to  m  do
```

begin

  if C_i is an accepting configuration **then** accept;

  let C_i^{yes} and C_i^{no} be the follow configurations

  of C_i and let y be the query string ;

  if $y \in A$ **then** list := list $\cup \{C_i^{yes}\}$

      else list := list $\cup \{C_i^{no}\}$;

end; {for}

end; {while}

reject ; {no accepting configuration found and no configuration left to explore} .

end.

leave the verification to the reader. □

Book, Long and Selman (1984) also prove similar positive rela-vization results for the questions NP =? co-NP and PH =? PSPACE.

rollary 6.4. (Long and Selman, 1983)

P = NP if and only if for every tally set T, P(T) = NP(T).

oof. Every nondeterministic, polynomial time bounded oracle machine at asks the oracle only for strings over a one-letter alphabet ivially satisfies Definition 6.2. □

Thus, $P \neq NP$ if a tally set T can be constructed such that P(T) NP(T). Notice that the set B constructed in Theorem 6.1 is not too r from being tally, it is sparse. For other comparisons of classes, sitive relativizations using sparse sets are known (Long and Selman, 83; Dowd, 1982). The following result is from (Long and Selman, 1983). e proof method stems from (Balcazar, Book and Schoning, 1984).

eorem 6.5

(i) $PH = \Sigma_k^p$ iff for each sparse set S, $PH(S) = \Sigma_k^p(S)$ $(k \geq 2)$.

(ii) PSPACE = PH iff for each sparse set S, PSPACE(S) = PH(S).

(iii) $PP \subseteq PH$ iff for each sparse set S, $PP(S) \subseteq PH(S)$.

oof. The non-trivial proof direction is always from left to right.

(i) Let $L \in \Pi_k^P(S)$ for a sparse set S, and suppose that $PH = \Sigma_k^P$ (thus, $\Sigma_k^P = \Pi_k^P$). It suffices to show that $L \in \Sigma_k^P(S)$. By the quantifier characterization of the (relativized) polynomial hierarchy,

$$L = \{ x \mid (\forall y_1)_p (\exists y_2)_p \ldots (Q_k y_k)_p \ \langle x, y_1, \ldots, y_k \rangle \in L(M, S) \}$$

for a polynomial p and a deterministic, polynomial time bounded oracle machine M. There is a polynomial q such that M on inputs of the form $\langle x, y_1, \ldots, y_k \rangle$, $|x| = n$, $|y_1|, \ldots, |y_k| \leq p(n)$, does not query the oracle for strings longer than $q(n)$. By the fact that S is sparse, there is a polynomial r such that $|S_{\leq n}| \leq r(n)$ for all n. Now we give a characterization of L similar to the proof of Lemma 5.7.

$L = \{ x \mid (\exists w)$ [w encodes a set of at most $r(q(|x|))$ strings of size at most $q(|x|)$ and
$(\forall u)_q$ [$u \in S$ iff u is in the set encoded by w]
and $\langle x, w \rangle \in D$] $\}$.

Observe that the first existential quantifier can be bounded by a suitable polynomial. The set D occuring above is the following.

$D = \{ \langle x, w \rangle \mid (\forall y_1)_p (\exists y_2)_p \ldots (Q_k y_k)_p$ [M accepts the input $\langle x, y_1, \ldots, y_k \rangle$ where oracle queries for some string y are simulated with the answer "yes" iff y is in the set encoded by w] $\}$.

Clearly, $D \in \Pi_k^P$. By the assumption $\Sigma_k^P = \Pi_k^P$, it follows $D \in \Sigma_k^P$. Therefore, L has a characterization of the form

$$(\exists w) \ [\ S_1(w) \ \text{and} \ (\forall u) \ S_2(u, w) \ \text{and} \ D(x, w)\ ]$$

where $S_1 \in P$, $S_2 \in P(S)$, and $D \in \Sigma_k^P$. For $k \geq 2$ it follows $L \in \Sigma_k^P(S)$.

(ii) The proof is similar to (i). The appropriate set D is now in PSPACE. From the assumption PSPACE = PH it then follows $D \in \Sigma_k^P$ for some $k \geq 0$.

(iii) Similar to (i). The set D is now in PP, and by the

ssumption, D is in PH. □

.2. NEGATIVE RELATIVIZATIONS

As argued above, a "negative" relativization result seems to be more lesirable. For the P-NP-question this seems very hard to achieve. Our esults apply to the classes of the polynomial-time hierarchy.

heorem 6.6. (Balcázar, Book and Schöning, 1984c)

(i) The polynomial hierarchy is infinite if and only if it is nfinite relative to every sparse oracle.

(ii) PH ≠ PSPACE if and only if for every sparse set S, PH(S) ≠ SPACE(S).

(iii) PP $\not\subseteq$ PSPACE if and only if for every sparse set S, PP(S) $\not\subseteq$ H(S).

roof. The trivial proof direction is always from right to left.

(i) Suppose there is a sparse set S such that the polynomial ierarchy is finite relative to S, that is, for some $k \geq 0$, PH(S) = ^p_k(S). Now consider QBF_{k+1}, a self-reducible and Σ^p_{k+1}-complete set cf. section 4.5). Then we have

$$QBF_{k+1} \in \Sigma^p_{k+1} \subseteq PH \subseteq PH(S) = \Sigma^p_k(S) \subseteq \Sigma^p_k/Poly$$

sing Theorem 4.21. By Theorem 4.27,

$$\Sigma^p_{k+3} = \Sigma^p_2(QBF_{k+1}) = \Sigma^p_{k+3} \ .$$

hat means, the polynomial hierarchy collapses to Σ^p_{k+2}.

(ii) The proof is similar to (i). From the the assumption PH(S) PSPACE(S) follows that PSPACE = $\Sigma^p_k(S) \subseteq \Sigma^p_k/Poly$ for some $k \geq 0$. hen by Corollary 4.29, PSPACE = Σ^p_{k+2} = PH.

(iii) Similar to (i) and (ii). We need to show the existence of a et that is both self-reducible and PP-complete. Then Theorem 4.27

applies. The set #SAT is known to be PP-complete (cf. section 3.1). The following lemma completes the proof of (iii).

<u>Lemma 6.7</u>. The set #SAT = { <i, F> | F has at least i different satisfying assignments } is self-reducible.

<u>Proof</u>. For each Boolean formula F let F_a, $a \in \{0,1\}$, be the formula obtained from F by replacing the first occuring variable in F by the constant a. The encoding of F_a is assumed to be shorter than the encoding of F. Let #F denote the number of assignments (to the variables occuring in F) that make F true. Observe that for each Boolean formula F, $\#F = \#F_0 + \#F_1$. Using this observation, we can define a self-reducing oracle machine for #SAT.

```
begin
  input <i, F> ;
  if i = 0 then accept ;
  if F contains no variables
    then
      if F is true and i = 1 then accept
                                 else reject
    else
      begin
        using binary search, find the largest
        k ∈ {0,...,i} such that <k, F0> ∈ #SAT ;
        using binary search, find the largest
        n ∈ {0,...,i} such that <n, F1> ∈ #SAT ;
        if i ≤ k+n then accept
                   else reject ;
      end ;
end.
```

The correctness of the algorithm is easy to see using the above observation. For the run time of this algorithm, note that a binary search in an interval of size m needs time O(log m), and in this case we have $m \leq 2^n$, n = number of variables in F. Hence the machine is polynomially time bounded and #SAT is self- reducible. □

The proof of Theorem 6.6 (iii) is now completed. Observe that this is a non-trivial example of a self-reduction. In the cases of SAT, QBF, QBF_k, the self-reduction is of the simpler "2-questions-truth-table" type. In (Simon, 1980) a weaker <u>nondeterministic</u> self-reduction for #SAT is introduced. (Guess the numbers k and n and verify using two oracle queries).

We summarize the results from Theorems 6.5 and 6.6 in a corollary. (cf. Balcázar, Book, Long, Schöning and Selman, 1984).

<u>Corollary 6.8</u>.

(i) The polynomial hierarchy is finite if and only if the polynomial hierarchy is finite relative to <u>all</u> sparse oracles if and only if there <u>exists</u> a sparse oracle relative to which the polynomial hierarchy is finite.

(ii) PH = PSPACE if and only if for <u>all</u> sparse oracles S, $PH(S) = PSPACE(S)$ if and only if there <u>exists</u> a sparse oracle S such that $PH(S) = PSPACE(S)$.

(iii) $PP \subseteq PH$ if and only if for <u>all</u> sparse oracles S, $PP(S) \subseteq PH(S)$ if and only if there <u>exists</u> a sparse oracle set S such that $PP(S) \subseteq PP(S)$.

That is, the above three open questions (i) whether the polynomial hierarchy is finite, (ii) whether PH = PSPACE, and (iii) whether $PP \subseteq PH$ are (positively or negatively) answered if and only if they can be answered in the presence of a sparse oracle. E.g., there cannot exist two sparse sets S_1 and S_2 such that $PH(S_1) = PSPACE(S_1)$, but $PH(S_2) \neq PSPACE(S_2)$.

In interesting contrast to these result concerning the polynomial hierarchy is the existence of a sparse oracle that separates P and NP given in Theorem 6.1.

6.3. HIGHER LEVELS OF THE POLYNOMIAL HIERARCHY

In the original definition, the polynomial hierarchy is built up to level k using relativized NP-computations with oracles from level k-1, that is, $\Sigma_k^p = NP(\Sigma_{k-1}^p)$. We want to study the question how these relativized computations look like if k is larger ($k \geq 2$). First we prove that a non-polynomial number of oracle queries is necessary to obtain a set in $\Sigma_k^p - \Delta_k^p$ ($k \geq 2$).

__Theorem 6.9__. (Book, Long and Selman, 1984) Let $A \in \Sigma_k^p - \Delta_k^p$ ($k \geq 2$), and let $A = L(M, B)$ for a set B in Δ_k^p and for a nondeterministic, polynomial time bounded oracle machine M. Then for all polynomials p and infinitely many inputs x, M asks more than $p(|x|)$ many questions to the oracle B in the nondeterministic computation tree induced by M, x, and B. (In other words, M is not a machine in the sense of Definition 6.2).

__Proof__. Suppose to the contrary that $A \in NP_b(B)$. Consider the proof of Theorem 6.3. There, a set $H \in NP$ is defined such that $A \in P(H \oplus B)$. Thus, $A \in P(\Delta_k^p) = \Delta_k^p$, contradicting the assumption that $A \notin \Delta_k^p$. □

__Corollary 6.10__. (Long, 1985) Let $A \in \Sigma_k^p - \Delta_k^p$ ($k \geq 2$), and let B be in Σ_{k-1}^p such that $A \in NP(B)$. Then B is not sparse.

__Proof__. Suppose $B \in \Sigma_{k-1}^p$ is sparse and $A \in NP(B)$. Then for C = prefix(B) (see Lemma 5.21) we have: $C \in \Sigma_{k-1}^p$, and $A \in NP_b(C)$ by the proof method of Lemma 5.21. This contradicts Theorem 6.3. □

Next we consider all "derivatives" of the set $A \in \Sigma_k^p$.

__Theorem 6.11__. (Balcázar, Book and Schöning, 1984ab) Let $A \in \Sigma_k^p - \Pi_k^p$ ($k \geq 3$), and let $D_1, \ldots, D_{k-1}$ be sets with $D_1 \in NP$, $D_i \in NP(D_{i-1})$ for $i = 2,\ldots,k-1$, and $A \in NP(D_{k-1})$. Then $D_1, \ldots, D_{k-2}$ do not have polynomial size circuits.

__Proof__. Let $i \in \{1,\ldots,k-2\}$, and suppose $D_i \in P/Poly$. By Lemma 5.6, there is a set $C \in CIR(D_i)$ in the class $\Pi_1^p(D_i) \subseteq \Pi_{i+1}^p$. By the fact that $A \in \Sigma_{k-i}^p(D_i)$, A can be characterized as

$$A = \{x \mid (\exists y_1)_p (\forall y_2)_p \ldots (Q_{k-i} y_{k-i})_p \langle x, y_1, y_2, \ldots, y_{k-i}\rangle \in L(M, D_i)\}$$

where p is some polynomial, and M is a deterministic, polynomial time bounded oracle machine. Let $C \in CIR(D_i)$ be witnessed by "circuit interpreter" B. Then, very similar to the proof of Lemma 5.7, A can be characterized as

$$A = \{ x \mid (\forall w) \ [\langle 0^{r(n)}, w \rangle \in C \ \rightarrow \ (\exists y_1)_p (\forall y_2)_p \dots (Q_{k-i} y_{k-i})_p \ \langle x, y_1, \dots, y_{k-i} \rangle \in L(M, B_w)] \}.$$

All quantifiers are polynomially bounded, and r is a suitable chosen polynomial (cf. the proof of Lemma 5.7). The form of this predicate shows that $A \in \Pi^P_{\max\{i+2,k-i+1\}} \subseteq \Pi^P_k$, contradicting the assumption that $A \notin \Pi^P_k$. □

REFERENCES

Adleman, L. (1978), Two theorems on random polynomial time, 19th IEEE Sympos. Foundations of Comput. Sci., 75-83.

Adleman, L. and Manders, K. (1977), Reducibility, randomness and intractability, Proc. 9th Annual ACM Sympos. Theory of Computing, 151-163.

Adleman L. and Manders, K. (1979), Reductions that lie, 20th IEEE Sypos. Foundations of Comput. Sci., 397-410.

Baker, T., Gill, J. and Solovay, R. (1975), Relativizations of the P=?NP question, SIAM Journal on Computing 4, 431-442.

Balcázar, J.L., Book, R.V. and Schöning, U. (1984a), Sparse oracles, lowness, and highness, 11. Intern. Sympos. Math. Foundations of Comput. Sci., Lecture Notes in Computer Science 176, 185-193, Springer-Verlag, Heidelberg.

Balcázar, J.L., Book, R.V. and Schöning, U. (1984b), Sparse sets, lowness and highness, preprint, to appear in SIAM Journal on Computing.

Balcázar, J.L., Book, R.V. and Schöning, U. (1984c), The polynomial-time hierarchy and sparse oracles, preprint, to appear in Journal of the ACM.

Balcázar, J.L., Book, R.V., Long, T.J., Schöning, U. and Selman, A.L. (1984), Sparse oracles and uniform complexity classes, 25th IEEE Sympos. Foundations of Computer Science, 308-311.

Balcázar, J.L. and Schöning, U. (1985), Bi-immune sets for complexity classes, Math. Systems Theory 18, 1-10.

Bennett, C.H. and Gill, J. (1981), Relative to a random oracle A, $P^A \neq NP^A \neq coNP^A$ with probability 1, SIAM Journal on Computing 10, 96-113.

Berman, P. (1978), Relationship between density and deterministic complexity of NP-complete languages, Symp. on Math. Foundations of Computer Science, Lecture Notes in Computer Science 62, 63-71, Springer-Verlag, Heidelberg.

Berman, L. and Harmanis, J. (1977), On isomorphism and density of NP and other complete sets, SIAM Journal on Computing 6, 305-327.

Book, R.V. (1974), Tally languages and complexity classes, Information and Control 26, 186-193.

Book, R.V., Long, T.J. and Selman, A.L. (1984), Quantitative relativizations of complexity classes, SIAM Journal on Computing 13, 461-487.

Chernoff, H. (1952), A measure of asymptotic efficiency for tests of a hypothesis based on a sum of observations, Ann. Math. Stat. 23, 493-509.

Cook, S.A. (1971), The complexity of theorem proving procedures, Proc. 3rd Ann. ACM Sympos. Theory of Computing, 151-158.

Dowd, M. (1982), Forcing and the P hierarchy, Technical Report LCSR-TR-35, Rutgers University, New Brunswick, New Jersey.

rdős, P. and Spencer, J. (1974), Probabilistic Methods in Combinatorics, Academic Press, New York.
ven, S., Selman, A.L. and Yacobi, Y. (1985), Hard-core theorems for complexity classes, Journal of the ACM 32, 205-217.
eller, W. (1957), An Introduction to Probability Theory and Its Applications, Volume 1, Wiley, New York.
ischer, M. (1974), Lectures on network complexity, Preprint, University of Frankfurt.
ortune, S. (1979), A note on sparse complete sets, SIAM Journal on Computing 8, 431-433.
arey, M.R. and Johnson, D.S. (1979), Computers and Intractability: A Guide to the Theory of NP-Completeness, Freeman, San Franscico.
ill, J. (1977), Computational complexity of probabilistic Turing machines, SIAM Journal on Computing 6, 675-695.
arrison, M.A. (1965), Introduction to Switching and Automata Theory, McGraw-Hill, New York.
artmanis, J. (1983a), On sparse sets in NP - P, Information Proc. Letters 16, 55-60.
artmanis, J. (1983b), On non-isomorphic NP complete sets, Technical Report TR 83-576, Cornell University, Ithaca, NY.
artmanis, J. and Hopcroft, J.E. (1976), Independence results in Computer Science, SIGACT News 8, 4, 13-23.
artmanis, J., Immerman, N. and Sewelson, V. (1983), Sparse sets in NP-P: EXPTIME versus NEXPTIME, Proc. 15th ACM Symp. Theory of Computing, 382-391.
artmanis, J. and Yesha, Y. (1983), Computation times of NP sets of different densities, Automata, Languages and Programming, Lecture Notes in Computer Science 154, 319-330, Springer-Verlag, Heidelberg.
opcroft, J.E. and Ullman, J.D. (1979), Introduction to Automata Theory, Languages, and Computation, Addison-Wesley, Reading, Mass.
eller, H. (1983), On relativized polynomial hierarchies extending two levels, Conference on Computational Complexity Theory, Santa Barbara, 109-114.
okusch, C. (1968), Semirecursive sets and positive reducibility, Transactions of the AMS 131, 420-436.
annan, R. (1982), Circuit-size lower bounds and nonreducibility to sparse sets, Information and Control 55, 40-56.
arp, R.M. (1972), Reducibility among combinatorial problems, in: Miller, Thatcher (ed.), Complexity of Computer Computations, Plenum Press, New York, 302-309.
arp, R.M. and Lipton, R.J. (1980), Some connections between nonuniform and uniform complexity classes, Proc. 12th ACM Symp. Theory of Computing, 302-309.
o, K. (1982), Some observations on the probabilistic algorithms and NP-hard problems, Information Processing Letters 14, 39-43.
o, K. (1983), On self-reducibility and weak p-selectivity, Journal of Computer ans System Sciences 26, 209-221.
o, K. and Moore, D. (1981), Completeness, approximation and density, SIAM Journal on Computing 10, 787-796.
o, K. and Schöning, U. (1985), On circuit-size complexity and the low hierarchy in NP, SIAM Journal on Computing 14, 41-51.

Ladner, R.E. (1975a), The circuit value problem is log space complete for P, SIGACT News 7, 18-20.

Ladner, R.E. (1975b), On the structure of polynomial time reducibilities, Journal of the ACM 22, 155-171.

Lautemann, C. (1983), BPP and the polynomial hierarchy, Information Processing Letters 17, 215-217.

Lerman, M. (1983), Degrees of Unsolvability, Springer-Verlag, Heidelberg

Long, T.J. (1982a), A note on sparse oracles for NP, Journal of Computer and System Sciences 24, 224-232.

Long, T.J. (1982b), Strong nondeterministic polynomial-time reducibilities, Theoretical Computer Science 21, 1-25.

Long, T.J. (1985), On restricting the size of oracles compared with restricting access to oracles, SIAM Journal on Computing 14, 585-597.

Long, T.J. and Selman, A.L. (1983), Relativizing complexity classes with sparse oracles, preprint, to appear in Journal of the ACM.

Lupanov, O.B. (1958), On the synthesis of contact networks, Dokl. Akad. Nauk. SSSR 119, No. 1, 23-26.

Lynch, N. (1975), On reducibility to complex or sparse sets, Journal of the ACM 22, 341-345.

Mahaney, S. (1982), Sparse complete sets for NP: solution of a conjecture of Berman and Hartmanis, Journal of Computer and System Sciences 25, 130-143.

Meyer, A. and Paterson, M. (1979), Whit what frequency are apparently intractable problems difficult?, MIT/LCS/TM-126, Lab. for Computer Science, MIT, Cambridge, Mass.

Orponen, P. and Schöning, U. (1984), The structure of polynomial complexity cores, 11th Symp. on Math. Foundations of Computer Science, Lecture Notes in Computer Science 176,, 452-458.

Orponen, P. and Schöning, U. (1985), The density and complexity of polynomial cores for intractable sets, submitted for publication.

Pippenger, N. (1979), On simultaneous resource bounds, 20th IEEE Symp. Foundations of Computer Science, 307-311.

Rabin, M.O. (1976), Probabilistic algorithms, in: Traub, J.F. (ed.), Algorithms and Complexity, Academic Press, New York.

Savage, J.E. (1976), The Complexity of Computing, Wiley, New York.

Schnorr, C.P. (1976), Optimal algorithms for self-reducible problems, 3r Int. Colloq. Automata, Languages and Programming, Edinburgh, University Press, 322-337.

Schöning, U. (1982), A uniform approach to obtain diagonal sets in complexity classes, Theor. Computer Science 18, 95-103.

Schöning, U. (1983), A low and a high hierarchy within NP, Journal of Computer and System Sciences 27, 14-28.

Schöning, U. (1984), On small generators, Theor. Computer Science 34, 337-341.

Selman, A.L. (1978), Polynomial time enumeration reducibility, Theor. Computer Science 14, 91-101.

Selman, A.L. (1979), P-selective sets, tally languages, and the behavior of polynomial time reducibilities on NP, Math. Systems Theory 13, 55-65.

Selman, A.L. (1982a), Analogues of semirecursive sets and effective reducibilities to the study of NP complexity, Information and Contro 52, 36-51.

Selman, A.L. (1982b), Reductions on NP and p-selective sets, Theor. Computer Science 19, 287-304.

Selman, A.L. (1983), Remarks about natural self-reducible sets in NP and public-key cryptosystems, preprint.

Shannon, C. (1949), The synthesis of two-terminal switching circuits, BSTJ 28, 59-98.

Simon, J. (1975), On some central problems in computational complexity, Ph.D. dissertation, Cornell University, Ithaca, NY.

Simon, J. (1980), A note on sparse sets and probabilistic polynomial time, Technical report CS-80-13, Penn. State University, University Park, Pennsylvania.

Sipser, M. (1982), On relativization and the existence of complete sets, 9th Int. Colloq. Automata, Languages and Programming, Lecture Notes in Computer Science 140, 523-531, Springer-Verlag, Heidelberg.

Sipser, M. (1983), A complexity theoretic approach to randomness, Proc. 15th Ann. ACM Symp. Theory of Computing, 330-335.

Solovay, R. Strassen, V. (1977), A fast Monte-Carlo test for primality, SIAM Journal on Computing 6,84-85.

Stockmeyer, L.J. (1977), The polynomial-time hierarchy, Theor. Computer Science 3, 1-22.

Ukkonen, E. (1983), Two results on polynomial-time truth-table reductions to sparse sets, SIAM Journal on Computing 12, 580-587.

Valiant, L.G. (1976), Relative complexity of checking and evaluating, Information Processing Letters 5, 20-23.

Wilson, C.B. (1983), Relativized circuit complexity, 24th IEEE Symp. Foundations of Computer Science, 329-334.

Wrathall, C. (1977), Complete sets and the polynomial-time hierarchy, Theor. Computer Science 3, 23-33.

Yap, C.K. (1983), Some consequences of non-uniform conditions on uniform classes, Theor. Computer Science 26, 287-300.

Yesha, Y. (1983), On certain polynomial-time truth-table reducibilities of complete sets to sparse sets, SIAM Journal on Computing 12, 411-425.

Young, P. (1983), Some structural properties of polynomial reducibilities and sets in NP, Proc. 15th Ann. ACM Symp. Theory of Computing, 392-401.

Zachos, S. (1982), Robustness of probabilistic computational complexity classes under definitional perturbations, Information and Control 54, 143-154.

Zachos, S. and Heller, H.(1984), A decisive characterization of BPP, preprint.

SYMBOLS

INDEX

Vol. 167: International Symposium on Programming. Proceedings, 1984. Edited by C. Girault and M. Paul. VI, 262 pages. 1984.

Vol. 168: Methods and Tools for Computer Integrated Manufacturing. Edited by R. Dillmann and U. Rembold. XVI, 528 pages. 1984.

Vol. 169: Ch. Ronse, Feedback Shift Registers. II, 1–2, 145 pages. 1984.

Vol. 171: Logic and Machines: Decision Problems and Complexity. Proceedings, 1983. Edited by E. Börger, G. Hasenjaeger and D. Rödding. VI, 456 pages. 1984.

Vol. 172: Automata, Languages and Programming. Proceedings, 1984. Edited by J. Paredaens. VIII, 527 pages. 1984.

Vol. 173: Semantics of Data Types. Proceedings, 1984. Edited by G. Kahn, D. B. MacQueen and G. Plotkin. VI, 391 pages. 1984.

Vol. 174: EUROSAM 84. Proceedings, 1984. Edited by J. Fitch. XI, 396 pages. 1984.

Vol. 175: A. Thayse, P-Functions and Boolean Matrix Factorization. VII, 248 pages. 1984.

Vol. 176: Mathematical Foundations of Computer Science 1984. Proceedings, 1984. Edited by M.P. Chytil and V. Koubek. XI, 581 pages. 1984.

Vol. 177: Programming Languages and Their Definition. Edited by C. B. Jones. XXXII, 254 pages. 1984.

Vol. 178: Readings on Cognitive Ergonomics – Mind and Computers. Proceedings, 1984. Edited by G. C. van der Veer, M. J. Tauber, T. R. G. Green and P. Gorny. VI, 269 pages. 1984.

Vol. 179: V. Pan, How to Multiply Matrices Faster. XI, 212 pages. 1984.

Vol. 180: Ada Software Tools Interfaces. Proceedings, 1983. Edited by P. J. L. Wallis. III, 164 pages. 1984.

Vol. 181: Foundations of Software Technology and Theoretical Computer Science. Proceedings, 1984. Edited by M. Joseph and R. Shyamasundar. VIII, 468 pages. 1984.

Vol. 182: STACS 85. 2nd Annual Symposium on Theoretical Aspects of Computer Science. Proceedings, 1985. Edited by K. Mehlhorn. VII, 374 pages. 1985.

Vol. 183: The Munich Project CIP. Volume I: The Wide Spectrum Language CIP-L. By the CIP Language Group. XI, 275 pages. 1985.

Vol. 184: Local Area Networks: An Advanced Course. Proceedings, 1983. Edited by D. Hutchison, J. Mariani and D. Shepherd. VIII, 497 pages. 1985.

Vol. 185: Mathematical Foundations of Software Development. Proceedings, 1985. Volume 1: Colloquium on Trees in Algebra and Programming (CAAP'85). Edited by H. Ehrig, C. Floyd, M. Nivat and J. Thatcher. XIV, 418 pages. 1985.

Vol. 186: Formal Methods and Software Development. Proceedings, 1985. Volume 2: Colloquium on Software Engineering (CSE). Edited by H. Ehrig, C. Floyd, M. Nivat and J. Thatcher. XIV, 455 pages. 1985.

Vol. 187: F.S. Chaghaghi, Time Series Package (TSPACK). III, 305 pages. 1985.

Vol. 188: Advances in Petri Nets 1984. Edited by G. Rozenberg with the cooperation of H. Genrich and G. Roucairol. VII, 467 pages. 1985.

Vol. 189: M. S. Sherman, Paragon. XI, 376 pages. 1985.

Vol. 190: M. W. Alford, J. P. Ansart, G. Hommel, L. Lamport, B. Liskov, G. P. Mullery and F. B. Schneider, Distributed Systems. Edited by M. Paul and H. J. Siegert. VI, 573 pages. 1985.

Vol. 191: H. Barringer, A Survey of Verification Techniques for Parallel Programs. VI, 115 pages. 1985.

Vol. 192: Automata on Infinite Words. Proceedings, 1984. Edited by M. Nivat and D. Perrin. V, 216 pages. 1985.

Vol. 193: Logics of Programs. Proceedings, 1985. Edited by R. Parikh. VI, 424 pages. 1985.

Vol. 194: Automata, Languages and Programming. Proceedings 1985. Edited by W. Brauer. IX, 520 pages. 1985.

Vol. 195: H. J. Stüttgen, A Hierarchical Associative Processin System. XII, 273 pages. 1985.

Vol. 196: Advances in Cryptology: Proceedings 1984. Edited b G. R. Blakley and D. Chaum. IX, 491 pages. 1985.

Vol. 197: Seminar on Concurrency. Proceedings, 1984. Edited b S. D. Brookes, A. W. Roscoe and G. Winskel. X, 523 pages. 1985.

Vol. 198: A. Businger, PORTAL Language Description. VII 186 pages. 1985.

Vol. 199: Fundamentals of Computation Theory. Proceedings, 198 Edited by L. Budach. XII, 533 pages. 1985.

Vol. 200: J. L. A. van de Snepscheut, Trace Theory and VLSI Design VI, 0–140 pages. 1985.

Vol. 201: Functional Programming Languages and Computer Archi tecture. Proceedings, 1985. Edited by J.-P. Jouannaud. VI, 413 pages 1985.

Vol. 202: Rewriting Techniques and Applications. Edited b J.-P. Jouannaud. VI, 441 pages. 1985.

Vol. 203: EUROCAL '85. Proceedings Vol. 1, 1985. Edited b B. Buchberger. V, 233 pages. 1985.

Vol. 204: EUROCAL '85. Proceedings Vol. 2, 1985. Edited b B. F. Caviness. XVI, 650 pages. 1985.

Vol. 205: P. Klint, A Study in String Processing Languages. VIII, 16 pages. 1985.

Vol. 206: Foundations of Software Technology and Theoretical Com puter Science. Proceedings, 1985. Edited by S. N. Maheshwari. IX 522 pages. 1985.

Vol. 207: The Analysis of Concurrent Systems. Proceedings, 198 Edited by B. T. Denvir, W. T. Harwood, M. I. Jackson and M. J. Wray VII, 398 pages. 1985.

Vol. 208: Computation Theory. Proceedings, 1984. Edited by A Skowron. VII, 397 pages. 1985.

Vol. 209: Advances in Cryptology. Proceedings, 1984. Edited by T Beth, N. Cot and I. Ingemarsson. VII, 491 pages. 1985.

Vol. 210: STACS 86. Proceedings, 1986. Edited by B. Monien and G Vidal-Naquet. IX, 368 pages. 1986.

Vol. 211: U. Schöning, Complexity and Structure. V, 99 pages. 1986.